From The Flying Doctors
in the basement of No. 9

with many thanks

Christmas 1973

Power Play

The Tumultuous World of
Middle East Oil 1890-1973

Power Play

The Tumultuous World of Middle East Oil 1890-1973

Leonard Mosley

Weidenfeld and Nicolson

ISBN 0 297 76533 7

Printed in Great Britain
by Willmer Brothers Limited, Birkenhead

Contents

Foreword

Ever since man can remember, the cap of earth and water covering the area nowadays known as the Middle East has been leaking on to the surface of its deserts, mountains and seas the bitumen, natural gas and oily scum that betray the presence of petroleum underneath. There are accounts in ancient documents of fiery outcrops, gaseous eructations or bitumenous seepages all the way from the Caspian Sea in the north to the Persian Gulf in the south, and from the mountains of the Bakhtiaris in the east to the Egypto-Libyan desert in the west. The Eternal Fires of Iraq have been burning for five thousand years and – as will be seen in this narrative – have been doused only once in that time. The Egyptians embalmed their dead pharaohs with the aid of petroleum distillates brought from the Red Sea and the Western Desert. The Hanging Gardens of Babylon were held together by pitch between the bricks – and it still sticks the ruins together today. Wandering bedouin used bitumen to waterproof their tents from an outcrop lake at Burgan, near Kuwait, whose whereabouts was passed down through countless generations. And across the Caspian Sea, on the Apscheron Peninsula, in the Caucasus, Persians and Parsees came to worship in hundreds of fire temples which were lit by a myriad of flames known as the Fires of Zoroaster.

In the Bible, in Pliny, in Herodotus, and in the records of many another ancient writer and traveller in the region, there are references to the presence and to the use of oil in multifarious forms, as an object of worship, as light, as fuel, as building material, as a massage for weary limbs, as an aphrodisiac for enfeebled lovers, as an ointment for blemished skins, and as a medicine for internal aches and distresses.

In most of the places where biblical man dug pitch or worshipped fire there are petroleum wells today exploiting the oil beneath them. The big oil companies have, of course, done even better than the ancients, for they have found oil where the tribes never suspected it existed – for instance, beneath the waters of the Persian Gulf, and below the sandy wilderness of the Rub al Khali (the Empty Quarter) of Saudi Arabia, which only a handful of Arabs had ever visited until US oil technicians took them there.

From all these widely scattered fields, both old and newly discovered, no less than 24,260,000 barrels of crude oil are siphoned out of the earth beneath the Middle East *every day*. That is enough to satisfy the needs of Western Europe, Japan, Australasia, East and South Africa and two thirds of non-Communist Asia. Altogether 1,213,000,000 tons of oil are shipped out of the area in a year,* which is more than is produced by the United States and its chief suppliers in the Caribbean (Venezuela, Colombia and Trinidad) put together. The amount of money this makes (a) for the US, British, Dutch, French and Japanese oil companies and (b) for the Arab and Iranian governments who have granted concessions is so huge that it sounds like a US budget deficit in a particularly bad year. I do not wish to confuse the reader with too many astronomical figures, but anyone with a portable computer will be able to work out that if a barrel of oil is now sold by Middle East producers at £1 (which was the price prevailing at the beginning of 1972) then a *daily* sum of £24,260,000 is being made in sales.

From these receipts the oil companies paid to the State of Kuwait alone £460,000,000 or $1,150,000,000 in taxes and royalties in 1970. Kuwait is only the third largest oil-producing country in the Middle East. It has a registered population of 733,000, less than a medium-sized city.

These are fabulous sums in any currency, and they have changed the face of the Middle East, physically and psychologically, as well as financially. Some cynics would say that, as a result, what were once backward lands have simply become rich backward lands, and that all the inflow of enormous wealth has done is to make Arabs and Iranians more arrogant, more feckless and more venal than ever. Others would reply that no Arab could possibly compete in arrogance, profligacy

* The figures are for 1970.

and venality with Western oilmen (and their governments) who have been exploiting them for well over half a century.

This book, however, is not written for cynics but for those who ask questions every time there is a Middle East oil crisis, and never get adequate answers. I hope they will find them in the course of this narrative.

For the purposes of this survey, I have included Libya as one of the Middle East oil countries even though, geographically, it is in North Africa. Its petroleum policies are closely tied to those of its Arab neighbours, and only too often what its volatile leader decides to do today will be emulated by the sheikhs of the Persian Gulf tomorrow. Other countries with which the survey deals are: Iran, Iraq, Bahrain, Kuwait, Saudi Arabia, Qatar, Oman and the Union of Arab Emirates.* I have visited all the countries mentioned above and talked with local rulers, leaders, ministers and technicians, as well as to American, European and Japanese oilmen working on the spot. I have also (in their own countries where possible, otherwise in their places of hiding or exile) discussed the future of Middle East oil with Arab and Iranian opponents of the *status quo*. In New York, San Francisco, London, Paris and the Hague I have talked with the spokesmen of the oil industry and made use of the expert knowledge and the archives which they most willingly offered. Finally I have spent some time with the so-called 'independent' operators, and talked over old times with veteran drillers, geologists and Arabists wherever they have retired after their labours in the mountains and deserts of the Middle East.†

These are some who have helped to give shape and colour to the picture. Let me now begin to sketch it in.

---

* A federation of Abu Dhabi, Dubai, Sharjah, Al Fujairah, Ajmar, Um al Khawain and Ras al Khaimah, a parcel of ex-British protectorates along the Persian Gulf.

† A full list of acknowledgements will be found at the end of this book.

PART ONE

The Concessionaires

1 · The Millionaires of Baku

The first oil concessionaires in the Middle East were not British, American, French or Dutch but Armenians and Persian Tartars. They exploited the great field of petroleum lying beneath the eastern shores of the Caspian Sea, and in bringing it to the surface they turned the scrublands of Azerbaijan into a black desert, polluted the rivers and the sea, demeaned and degraded their workers, and sowed the seeds of the Russian Revolution. They also kept half of Europe's lamps alight, and they broke the Standard Oil Company's stranglehold on the world paraffin market.

The great oilfields of Baku are today a tightly run Soviet Russian monopoly with next to no liaison with the great petroleum industries to the south of them. But in the nineteenth and early part of the twentieth century they were part of the Middle East free-for-all. Baku and its Azerbaijan hinterland had belonged to Persia until 1813, but even after the Russians took it over it remained Persian in spirit, and only the ruthless domination of the Cossack armies kept it under control. The area was all but awash with oil. Along the Caspian shores, and particularly on the Apscheron Peninsula, the flat scrublands covered a vast reservoir of petroleum and gas. The peninsula was dotted with flaring gas fires and small Zoroastrian temples to which pilgrims came to worship from India. If you rowed five miles into the Caspian and threw a tow of lighted rope overboard the sea would catch alight from the gas bubbling up from the sea bed. For generations the natives had painstakingly dug wells by hand, filled leather bags with the seeping oil, and transported it to market for sale as fuel or lubricant. Then in the 1870s someone brought to the Caucasus the first steam drill. It was the same kind as those which were transforming the US oil industry and making a multi-millionaire out of John D. Rockefeller, and a team of Pennsylvanian drillers came with it. They sent the metal bit spinning into the soft, sandy soil and up came what was known locally as 'a spouter'. Oil had been struck at 124 feet. From that moment the rush was on.

Always on the lookout for sources of revenue, the Czarist government in St Petersburg was quick to exploit the bonanza. The viceroy received urgent orders to seize most of the oil-productive land and lease concessions to the highest bidders. A great auction was held in which most of the rich Armenians from Tiflis, Batum and Istanbul bid wildly against each other, and against the local Azerbaijani and Tartar khans, for the right to exploit 'spouters' of their own. They bid half a million pounds for land which had previously been valued at one hundredth of that sum, and then went off excitedly to erect derricks and start drillings. Since the land which had been auctioned off was confined to those areas where hand-dug wells had previously been in operation, they did not need to be told where to start. On the other hand, a young Swede named Robert Nobel, who had been in the Caspian area investigating a timber scheme, quietly acquired a strip of territory away from the coveted areas, and paid only a thousand pounds for it. Then he called in his three brothers, Emil, the frail but shrewd young scholar, Ludwig, the arms manufacturer, and Alfred,* the chemist and inventor, and asked them for advice and financial help. The oilfield which they started in 1874 soon became the biggest and most prosperous in Baku, with refineries to purify the oil and pipelines to take it to the railhead.

But others made fortunes in Baku too. One of the most flamboyant of the oil millionaires was an Armenian named Alexander Mantachoff, whose wild parties lasted for days, and whose appetite for food, wine, women and violence was notorious. A contemporary picture of him, at a meeting of the local owners' association, shows him dressed in the formal morning dress of the period, topped by a black silk cloak with what appears to be white lining; and from beneath his trousers peep highly-polished and spurred riding boots. He

* Better known for his invention of dynamite and the Nobel prizes which are named after him.

is carrying a whip, and the expression on his heavy bearded face is both arrogant and challenging. He prided himself on being an insatiable womanizer and an unbeatable fighter with knife, whip or his fists. He had no guilt about his excessive indulgences, and his favourite remark was: 'Only the weak are good, because they are not strong enough to be bad'.

So far as Baku was concerned, it was true. The weak were the native Tartar and Georgian villagers who worked for the oil concessionaires, and they lived and worked under particularly degrading conditions. Their living quarters were a series of wooden-hutted compounds in an area about ten miles from Baku known as Black Town. It was close to the wells and at the mercy of every 'spouter' that came in, so that the streets and houses were constantly soaked with viscous oil, and the air was a mixed stench of petroleum and excrement. Pay was low and hours long. Food was provided from primitive canteens. Employees worked under practically the same conditions as black men in South Africa today, forbidden to have wives with them, their movements restricted. Their sexual needs were satisfied by a weekly visitation of a cartload of drabs brought in from the brothels of Tiflis and Batum, and what money they saved from their wages they were encouraged to spend in company bars on cheap wine and vodka. Any tendency to revolt against the conditions of work was brutally discouraged by private armies of Cossacks maintained by the well-owners, who rode among the rebels and cut them down with knouts and swords. Ringleaders were publicly flogged.

While the Baku oil industry was still in its heyday, and Alexander Mantachoff was the swashbuckling king of the Apscheron Peninsula, a young man arrived to visit him from Turkey. What he saw and heard and experienced was to make a lasting impression upon him, and, through him, on the future of the Middle East oil industry.

His name was Calouste Sarkis Gulbenkian and he was nineteen years old. Like his host, he was an Armenian. His father had interests in Baku oil and sold and exported paraffin from the markets of Constantinople, successfully enough to have sent his son to France to learn the language and to England for his general and technical education. Calouste Gulbenkian did not waste the money spent on him. Through formidable drive and a precocious intelligence, he earned good degrees in science and engineering at King's College, London, though he was still in his teens. As a reward, his father sent him on a trip first to Baku and the Apscheron Peninsula, and then on to Mosul, in what was then Turkish-controlled Mesopotamia. In Baku the young Armenian met the Nobel brothers and toured their oil installations, and was then passed on to his fellow Armenian. The burly, bearded and moustachioed Mantachoff, a veritable giant of a man, was at first disconcerted by his thin, soft-spoken, studious-looking guest, and after one day of showing him around his wells and refineries left it to his underlings to take Gulbenkian on the detailed tour of the outlying fields which the young man had requested. They came back and reported that he had made little or no verbal comment, even when confronted by the awesome sight of a fortuitous 'spouter', though he had scribbled copious notes.

But once invited to Mantachoff's ornate seaside palace, the youth demonstrated that he was much less introverted than appearances indicated. He savoured all the delights the household had to offer, and they were plenty. True, when one of his host's fabulous parties was in progress, he took no part in the barebacked horse-races, the wrestling matches and trials of strength in which the big Armenian and his Tartar guests liked to engage. But he seized whatever else was offered. He ate the rich imported foods and drank the fine wines and champagne. He relished and applauded the tumblers, magicians and bellydancers Mantachoff had brought in to amuse his guests.

It was a complete contrast to the austere existence which his family followed in Constantinople. Mantachoff introduced him to the pleasures of the flesh.

When Calouste Sarkis Gulbenkian returned to Constantinople, he wrote a long account of his trip which was published in 1892 under the title of *La Transcaucasie et la Peninsule d'Apcheron: Souvenirs d'un voyage.* It was a straightforward, serious study of a people and an industry, and well received. It turned out to be his hail and farewell to oil as a tangible, smellable commodity. He was to become a power in the oil world and, thanks to it, one of the richest men in the world. But for the rest of his life, the nearest he got to petroleum was in the tank of his car. He never visited an oilfield again. He did not need to. Baku, Apscheron and Mosul had taught him what oil was like, and Mantachoff what power and wealth an oil concession could bring him. That was all he needed to know, and from the age of nineteen onwards he set out to make his fortune out of oil – without actually getting the greasy stuff on his fingers.

6

2 · The Absentee Concessionaire

Unlike the millionaires of Baku, most of the Middle East concessionaires who came after them shared Gulbenkian's distaste and kept as far away as possible from the reek and touch of the precious commodity which made their fortunes. There was, for instance, William Knox D'Arcy. In the official history of the giant combine, British Petroleum, whose success grew out of his manipulations, Knox D'Arcy is described as 'the father of the entire oil industry of the Middle East'.[1] A French account of the origins of the petroleum industry gives a graphic account of his adventures in the wildernesses of South-West Persia, doggedly searching for oil with a prayer on his lips and a theodolite over his shoulder.

Wizened, his face withered by the torrid sun of the Persian deserts, [writes the author] worn out but still believing in God and his own idea, more and more on his knees in front of his crucifix, D'Arcy continued his search.[2]

Whereas, in fact, D'Arcy's feet never even touched Persian soil and he never saw the oilfields which made him the 'father' of Middle East oil. His only visit to the Middle East was a two-day excursion to Cairo and the Pyramids when his ship was passing through the Suez Canal on the way from Australia to England, and that happened long before he even dreamed of investing in oil. It is doubtful if he ever saw a barrel of the stuff in its crude state. He was an absentee concessionaire from start to finish.

On the other hand, it is almost certain that if the risk had not been taken by Knox D'Arcy in the spring of 1901, the exploitation of the oilfields of Persia would have been delayed for a

7

generation and the course of British imperial politics considerably altered. A banker by profession, William Knox D'Arcy was a huge, plump figure of a man with twinkling blue eyes and a walrus moustache, plus a beautiful wife, Nina, whom he hung with precious jewellery. He had made his millions out of the fabulous Morgan gold strike in Queensland, Australia, and he had moved to London to enjoy the fruits of his lucky investment, where he soon became the talk of the town, for the parties at his house in Grosvenor Square were fabulous. At one of them, Nellie Melba and Enrico Caruso, the two greatest opera singers of their time, sang for their supper.

D'Arcy seemed happy with his social success. But one night at dinner, a guest named Sir Henry Drummond Wolff, who had been British ambassador to Persia, listened with considerable interest when a fellow diner chaffed D'Arcy about his passion for motor cars and maintained that they would never take the place of the carriage-and-pair. To which D'Arcy replied: 'I think you are mistaken. One day I believe the motor car will revolutionize our means of transport. There will be horseless omnibuses to carry the people, and horseless carts to transport their goods. It is already happening in America, and soon it will happen in the rest of the civilized world.'

To which Drummond Wolff had interpolated: 'And if the world is suddenly full of horseless carriages, where will the oil come from to fuel them?'

'We must find and extract more of it from the ground,' D'Arcy said. 'The oil is there. We must dig for it.'

It was this vision of a new age of locomotion which made the Australian millionaire more than willing to talk to a young French archaeologist named Jacques de Morgan, who had been surveying potential oilfields in south-west Persia. De Morgan was a protégé of Drummond Wolff and it was he who effected an introduction between the two men. From it came an agreement on Knox D'Arcy's part to back negotiations with the Persian court for an oil concession, and to provide the money for operations once it was secured.

After considerable complications a deal was done, and on 28 May 1901 the Shah of Persia signed over to William Knox D'Arcy a sixty-year exclusive concession to 'obtain, develop, render suitable for trade, carry away and sell natural gas, petroleum, asphalt and esocerite through the whole extent of

the Persian Empire', with the exception of a crescent of territories along Persia's northern frontiers abutting Russia. As a
gesture of appeasement towards the Czarist government, which
had tried to get the concession for themselves, these areas were
reserved for Russian exploitation. But despite these exceptions,
D'Arcy had been accorded a vast territory in which to search,
find and develop: altogether 480,000 square miles. In return
the Persians demanded that he should form a company to
exploit the concession within two years of the signature of the
agreement. In this company the Persians would be given
20,000 £1 shares and 16 per cent of the profits, plus a further
£20,000 in cash. They also insisted on the appointment of an
'Imperial Commissioner' to safeguard their interests in the new
venture, and chose for the job a retired Persian army officer
named General Antoine Kitabgi. They did not appear to be
aware that Kitabgi was already a member of D'Arcy's syndicate.

A happy D'Arcy countersigned the agreement in London,
bought his wife a new tiara, and celebrated the event with a
huge party to which Lord Curzon, the British foreign secretary,
and many other government leaders, came. Then he waited for
the oil (and the profits) to flow in.

But it wasn't quite as easy as that.

First of all, they began drilling in the wrong place. Jacques
de Morgan had enthused over 'petroliferous evidences' in
the Bakhtiari country (where, in fact, oil was eventually
exploited), but the geologist D'Arcy sent out himself had
recommended an area some hundreds of miles to the west. It
certainly had oil, but the quality and the quantity were both
disappointing. Moreover, the conditions under which the field
parties began their work were daunting in the extreme. As his
field manager D'Arcy chose a hard-bitten Englishman named
G. B. Reynolds, who had drilled for oil in some remote places
in the Far East; and Reynolds in turn picked as his assistant a
wiry American ex-driller named Hiram Rosenplaenter, who
had worked in Texas and Mexico. But neither of them was
prepared for the dust, desolation and heat of the Persian desert.
The sun hurled heat down from the dark sky like flaming
thunderbolts. The temperature was one hundred and ten
degrees by seven in the morning and rose steadily throughout
the day. The local tribesmen, who acknowledged no authority

from such remote places as Teheran, rode threateningly into the tented encampment and demanded tribute. Since he had no force with which to repel them, Reynolds took them into his employ instead as encampment 'protectors' and bestowed plentiful baksheesh on their chiefs. But the guards could not be trusted, and their khans were 'as full of intrigue as a nightingale's egg is pregnant with melody', Reynolds reported; and at the same time that they stole from the camp, they left behind fevers and fleas that drove the men crazy.

Rosenplaenter had used Polish drilling crews to begin with, but they stopped work so often on the grounds that another religious holiday had come round that he first cabled home for a Roman Catholic calendar, and then gradually replaced them with Canadians and Americans. He began a new well at a godforsaken spot called Chiah Surkh.

By June, 1903, [the official record recounts] it had reached a depth of 1,292 feet where 'fine, dark-coloured sand rock' was encountered 'which cuts all our drills to pieces'. The drillers were suffering from the effects of bad water caused by the bloated dead bodies of thousands of grasshoppers which had chosen a watery grave in the Sekhuan stream, the only source of supply. To thirst and dysentery, heat-stroke was soon added. Everyone, even the Indian doctor, suffered from it. 'We are now five Europeans suffering from a touch of the sun. Stewart was yesterday in a critical condition . . . I have built two thick-walled stone houses for the drillers but even in these the temperature rises to 120 degrees Fahrenheit and, as the nights are hot, the men get very little rest and less sleep. In spite of this we are working night and day at the well.'[3]

Back in London, all D'Arcy had to worry about was money. He had spent £250,000 on getting the men and equipment out to Persia; he had had to pay off Drummond Wolff, the ex-British minister to Persia, who had got tired of waiting for his reward; and he was having trouble with his other erstwhile partners. Regretfully he cancelled the order for one of the latest and most expensive of the new motor cars to come on the market, a Rolls-Royce, and he gave his bank a block of Morgan gold shares against his mounting overdraft. He even contemplated giving up his private box at Ascot race course – except for the royal family, he was, at that time, the only racegoer to have one – but reluctantly decided that perhaps that would be going too far. But he cabled Reynolds and Rosenplaenter to press on. At last, in January 1904, oil came surging out of a second well Rosenplaenter had sunk, and it looked like a winner.

'Glorious news from Persia,' D'Arcy announced. 'It is the greatest help to me.'

But not for long. Two months later, the well ran dry. Rosenplaenter, sick with dysentery, and broken-hearted at the failure of his superhuman efforts, allowed himself to be shipped back to the United States. Reynolds stayed on to reconnoitre new terrain and to start drilling again. D'Arcy was by this time harassing him almost daily to find him some oil, 'before disaster engulfs us all'. He did not let Reynolds know that he was negotiating with an American syndicate (which eventually turned him down) and with the Rothschild interests in an attempt to find outside money. The negotiations with the Rothschilds took place in the spring sunshine along the Croisette at Cannes and aboard the Rothschild yacht floating in the harbour, over glasses of champagne and delicate snacks of caviar and pâté de foie gras, at a moment when in Persia Reynolds was working on despite a particularly weakening dysentery, and grief over the loss of his pet dog which had succumbed to heat stroke. It was the ostentatious background of D'Arcy's talks which turned the tides of fortune for him, however; they were reported back to the British government in London. Alarmed that D'Arcy might be planning to sell out to French interests (as indeed he was), they dispatched an emissary to see him at once with instructions to appeal to his patriotism and sense of duty towards the Empire. He was asked to return to London for a meeting at which British, rather than foreign, aid would be discussed. D'Arcy acceded with a theatrical display of willingness to sacrifice himself in the cause of king and country, though in truth he was losing little, for the Rothschilds were driving an exceedingly hard bargain with him in between the glasses of champagne.

It was a lucky moment for William Knox D'Arcy. Two years earlier, the Royal Navy, egged on by an unswerving believer in oil fuel, Admiral Lord ('Jackie') Fisher, had agreed to experiment with oil as the motive power for the dreadnoughts of His Majesty's fleets. It was at the height of the coal versus oil controversy. At Fisher's insistence, a navy cruiser, the *Hannibal*, was fitted with oil tanks as well as coal bunkers, and sailed out of Portsmouth harbour on a series of sea trials. For the first two hours of her voyage the ship surged forward under coal-fired boilers, only a light smoke trickling from her three stacks. But then the order was given to change to oil firing. Almost at once, though the speed remained constant, the ship was engulfed in great clouds of black smoke.

'You see!' cried the defenders of coal, choking delightedly on the acrid fumes.

Fisher was also choking but with anger. 'All I see is *sabotage*,' he replied. And he may have been right, for it turned out that the cruiser, in spite of meticulous instructions, had been fitted with old-fashioned vaporizers instead of special new burners. Naval friends of the big coal owners, who were fearful of losing one of their most important markets, were suspected of deliberately misinterpreting orders. They succeeded in delaying the changeover.

But in 1904, when Fisher moved to the top of active naval direction, as first sea lord, he crushed his opponents once and for all. An Oil Committee was set up to superintend the conversion from coal to oil for all vessels in the Royal Navy. There was, however, one danger. The tactical advantages of oil for naval use were by now accepted, but strategically, maintained E. G. Pretyman, a member of parliament who headed the Oil Committee, the British Navy would be placed at a great disadvantage: 'Whereas we possessed in the British Isles the best supply of the best steam coal in the world, a very small fraction of the oilfields of the world lay within the British Dominions, and even these were situated within very remote and distant regions.'

Therefore, if Knox D'Arcy had even a fifty-fifty chance of finding viable oilfields in Persia, they must not be allowed to fall under foreign control. The Admiralty had already given a contract to the Burma Oil Company, a British firm, to supply them with oil from their fields in Burma and Assam. When D'Arcy arrived in London from his conferences with the Rothschilds in Cannes he was told that, with the help and approval of the British government, Burma Oil was willing to bail him out of his difficulties and henceforth pay a share of the expenses of his Persian operations. It did not occur to D'Arcy (or to any of his new directors) to go out to Persia themselves with the good news and encourage their weary and dispirited but still indefatigable field manager, G. B. Reynolds. They simply sent him even more peremptory orders to find a viable field.

But the tie up with Burma Oil, and through that company with the British government, did remove one worry from Reynold's roster of woes. Local tribesmen had been getting increasingly hostile, and had taken to sabotaging equipment and stealing supplies when the amount of baksheesh they were being paid fell below expectations. This was 1905, and the heyday of the British Empire, when one man waving a Union Jack and threatening to bring in a gunboat, could cow a nation. Though Persia was an independent country, it sometimes almost came apart at the seams under the pressure for its control between Russia and Britain, a tug-of-war in which the Russians planted their teams in the north and the British in the south. The Shah complained, prodded by the Russians, when he heard that the terms of D'Arcy's concession were about to be altered to accommodate Burma Oil. Immediately the new syndicate sent a note to Whitehall hoping that they could count upon the support of the government 'in the event of any attempt by the Shah of Persia to cancel or interfere with the concession or on the part of Russia or any other neighbouring State to hamper the working of it.' The Foreign Office, as its reply, dispatched a carefully-worded missive which said nothing any foreign government could object to, but suggested everything from gunboats to armies.

I am to state [the under-secretary wrote] that it is obviously impossible that a definite statement should be made in regard to hypothetical contingencies but your clients or any other British Company formed to acquire and work the Concession can count on such support and protection as British subjects are always entitled to expect from His Majesty's Government.

That the note meant much more than it said is indicated by the fact that, shortly afterwards, and much to Reynolds' relief, a small contingent of troops from the British Indian Army was sent to protect his field operations from further molestation from the Bakhtiaris. The British government did not inform the Shah of this decision until the troops were actually there, and then, in an explanatory note, strongly suggested that he would be foolish to object to their presence on his territory.

The contingent was commanded by a young English lieutenant named Arnold Wilson, a sensitive young man of heroic good looks, whose subsequent distinguished career in British business and politics was to end on a note of gallant tragedy.* He set up his camp beside the oil company's rigs and soon became an enthusiastic admirer of Reynolds, who was, he wrote in his diary, 'dignified in negotiation, quick in action, and completely single-minded in his determination to find oil.'

But years had gone by, and he had still not found enough to justify the cost of the operations. The concession had been signed in 1901. It was now 1908. Even the Burma Oil Company and their British government sponsors were beginning to lose heart. As for D'Arcy, he was keening over his overdraft again and cancelling the option on his latest Rolls-Royce. In the post-Christmas hangovers of 1908, the members of the syndicate met in London to discuss the gloomy prospects and appear to have decided that they had had enough. It was just about this time, as the sun burst through the January rains in south-west Persia, that Reynolds spudded-in† a new well at a spot known as Masjid-i-Sulaiman (the Mosque of Solomon). One night when young Lieutenant Wilson went across to join Reynolds for their usual sundowner of whisky and chlorinated water, he found the oilman in the depths of depression. He had received a cable that morning, brought by the camel courier from the telegraph office in Baghdad, saying that 'funds were exhausted and the decision reached finally and irrevocably was that he was to cease work, dismiss the staff, dismantle anything worth the cost of transporting to the coast for reshipment, and come home.'‡

* Sir Arnold Wilson (he was knighted in 1920) became an oilman himself later and a director of the Anglo-Persian Oil Company. In the years before World War II he entered Parliament as a Tory and preached an Anglo-German alliance, strongly opposing those who wanted war with Germany. But once war began he waived his parliamentary immunity and at once (though fifty-five) volunteered for the RAF. He was commissioned as a rear-gunner in a bomber and was shot down and killed over Dunkirk by German ack-ack while bombing Nazi-held France.

† *Spudding-in* is the technical term for boring an oilwell.

‡ Oilmen have always cast doubt on this story, pointing to the fact that no copy of the cable exists in the archives of Burma Oil or British Petroleum. In view of what subsequently happened, someone may have quietly removed it from the files. Certainly there seems no reason to doubt Wilson, who had no motive for making up the story of the telegram.

Wilson was distressed at the emotional state of the older man, and indignant at 'the short-sighted decision which may result in the cancellation of the concession.'

I am tired of working here for these stay-at-home business men who, in all the years they have had the concession, have never come near it [Wilson wrote in his diary]. They have all the vices of absentee landlords. Cannot Government be moved to prevent these faint-hearted merchants, masquerading in top-hats as pioneers of Empire, from losing what may be a great asset?

But when he went to see Reynolds next day, the field manager had recovered his old spirit, and defiance had taken the place of despair. He was damned, he told Wilson, if he was going to take precipitate action on the strength of a cable. The drilling would go on. He sent a message to London to tell the syndicate of his decision, and received written permission to go on drilling until he reached 1,600 feet.

On 26 May 1908, oil was struck at Masjid-i-Sulaiman at 1,200 feet and a gusher came up over the top of the derrick. Joyfully, Reynolds called for a camel courier and sent off a message to the telegraph office in Baghdad to inform his head office of the good news. But young Wilson had telegraph facilities of his own, to the British resident in Bushire, and was anxious that the British government should know about the strike as urgently as possible. He realized that there would be a run on Burma Oil's shares as a result of the news, and therefore did not want any clerk along the wire to read his message and spread the news before his masters had seen it and taken appropriate action. He had no code that could take care of the situation, but he knew his prayer book, and rightly assumed that his minister would realize the significance of the message he dispatched and send it urgently on to London. It read: SEE PSALM 104 VERSE 15 THIRD SENTENCE AND PSALM 114 VERSE 8 SECOND SENTENCE. ['That he may bring out of the earth oil to make him a cheerful countenance ... the flint stone into a springing well.']

The strike at Masjid-i-Sulaiman proved to be the foundation of the syndicate's fortune, and no one expected to wax fatter on the glory of it all than William Knox D'Arcy. In April 1909 the Anglo-Persian Oil Company was formed to exploit this field and the remainder of the Persian concession. Its original capital of £1,000,000 was provided by Burma Oil and, at the suggestion of the British government, the choice of chairman of the company was left to the civil servants in Whitehall. To D'Arcy's fury, they passed him over in favour of an eighty-eight-year-old Tory peer named Lord Strathcona, through whom they hoped to keep a tight hold on the affairs of the new company. Then they awarded the Anglo-Persian a £20,000,000 contract to supply the Royal Navy with fuel oil.

In 1914, at the urgent insistence of Admiral Lord Fisher and his close friend, Winston Churchill, who sat in the cabinet as first lord of the Admiralty, they went even further. Thanks to Churchill's oratory, and dire warnings of Britain's danger of 'oil starvation' in the event of war with Germany, a bill was driven through parliament giving the government the right to take over control of Anglo-Persian Oil. Just a few months before the outbreak of World War I, the government bought 51 per cent of the company's shares for £2,200,000.

They gained control of a company (now named British Petroleum) which is capitalized today at £425,000,000 and had annual sales in 1970 of £2,600,000,000.

As for William Knox D'Arcy, he was soon complaining that he had been cheated out of his place in the hall of fame, and that the new controllers of Persian Oil were trying to conceal the fact that he was the father of it all. He died, an unhappy man, in 1917. And he never did get around to visiting the concession which changed the pattern of the world oil industry.

3 · Big Deal

So far the Americans had kept clear of the Middle East oil rush. They were too busy coping with the flow of oil from their own fields in Pennsylvania, California, Texas and Oklahoma Territory. But when news reached the US in the spring of 1908 that the D'Arcy concession had struck a rich petroleum field in Persia, several interests in New York decided that it was time to put a toe in the Middle East pool. Since the British and the Russians between them had parcelled up Persia, the obvious pool to dabble in was Iraq, where it had long been known that there were petroleum manifestations in the areas around Mosul and Kirkuk. The restive and reluctant Kurds and Arabs who populated these areas came under the control of the Turkish Empire in 1908, and it was therefore from Constantinople, the Turkish capital, that any concessions must be obtained. Constantinople at this time was ruled by the ineffable Sultan Abdul Hamid and was probably the most corrupt capital city in the world.

It so happened that an American concession-seeker was already in Turkey when news came of the D'Arcy oil strike. He was a retired rear-admiral in the US navy, a veteran of the sea battles of the Spanish-American War, and a former naval attaché in the US embassy to Turkey. His name was Colby Chester. He was not worried when other Americans appeared on the scene, nor was he concerned that other nations were haggling in the bazaars, for he was convinced that he already had the whole deal signed and sealed. His earlier service in Turkey had taught him that the way to get a bargain from the officials at Sultan Abdul Hamid's court was to keep your temper under all circumstances, and be ready with the bribe. His mission in 1908 had the backing of a consortium consisting of the New York Chamber of Commerce, the New York Board of

Trade and the New York Transport Association, and they had provided him with ample funds with which to suborn the venal officials of the Turkish court. In addition, thanks to friends in the Navy Department, he had good relations with the US government and could always rely on the aid and influence of the State Department.

With all this behind him, and thanks to a substantial bribe which he handed over to the grand vizier of the sultan's court, he was able in the spring of 1908 to emerge from Topkapi Palace brandishing not one concession but several, giving him the right to exploit oil and minerals in all parts of the Turkish Empire. Unfortunately, a few weeks after the concessions were signed the sultan lost his throne in the coup d'état of the Young Turks, and the privy purse territories to which Colby Chester had been given the rights reverted to the state. The concessions were worthless, and the admiral had to begin all over again. This time he had to contend with rivals of a rather more serious calibre and determination.

This was the moment when the Germany of Kaiser Wilhelm was challenging the might of the British Empire, the greatest power of the time, by proposing to build a railway all the way from Constantinople to Baghdad, where it would bring the Germans to the threshold of British India. As part of their project they had secured a concession from the Turks for a thousand-mile strip of territory through Turkey and Mesopotamia, and they not only had the right to construct a railway through it, but they had also acquired permission to exploit all oil and minerals to be found for twenty miles on either side of the line. True, these rights, like those of Colby Chester, were declared null and void when the Young Turks came to power, but since the new administration failed to reimburse the Germans for the $100,000 they had spent surveying the route, Berlin refused to accept the cancellation and maintained its rights under international law.

The British, alarmed at these activities on her imperial doorstep, decided that it was a moment when she too should interest herself in Turkish affairs. With the approval of the British government, a group of London financiers agreed to send one of their number, Sir Ernest Cassel, to Constantinople to look over the possibilities. Cassel could hold his own in the London banking world but realized that he would need a guide through the serpentine corridors of Turkish high finance. He picked Calouste Sarkis Gulbenkian as his adviser, and he could not have chosen a better.

Much had happened to the Armenian in the years since he had supped at Alexander Mantachoff's lavish table in the Baku oilfields in the 1890s. Gulbenkian was thirty-nine now, and had fined down his youthful enthusiasms to the pursuit of three objectives: oil concessions, money and women. There had been a time when all his energies had been engaged simply in the fight to survive. In the mid-nineties the Turks, like the Russians, had turned upon the Armenians in their midst and pillaged their houses, raped their women, and massacred old and young alike. Gulbenkian had succeeded in making his escape in a boat leaving for Egypt, and at the last moment good luck sailed with him. Milling among the panic-stricken crowds on the quayside who should he see but Mantachoff, who had come to Constantinople on an unheralded (and no doubt nefarious) visit and had found himself in the middle of a pogrom. Gulbenkian used his influence with the ship's owners, who were related to his in-laws, to get the oil millionaire aboard and forced his own young wife, who was pregnant, to quit her cabin in order to make Mantachoff comfortable. He was ever ready to do a service for a potential benefactor, and his eager servility towards the Armenian giant paid off. He became first Mantachoff's secretary, then the representative in London of his and various other Russian oil interests. He had never looked back. He had discovered his skill in complicated negotiations at a time when American oil interests (especially Standard Oil) were trying to cut-throat their way to control of the world fuel market. Gulbenkian was convinced that the only way for non-US oilmen to survive was consolidation, so that they could jointly meet the American competition with a united front. First he had brought the Russian oil interests into the big European organization, Royal Dutch. Then, when this rejuvenated company came under the control of the powerful Dutch operator, Henri Deterding, he had scurried back and forth as his emissary in the tortuous negotiations with the Shell Oil Company, out of which came the European consortium, Royal Dutch-Shell, a combination powerful enough to stand up to any price-cutting competition from Standard Oil. And Standard had called a truce.

Now Calouste Gulbenkian was well on the road to the eminence he planned for himself. He was a naturalized Briton, the father of a boy and a girl; he had houses in fashionable Hyde Park Terrace in London and the Boulevard Haussman in Paris. He also had an apartment in the Ritz Hotel, in Paris, where, for the rest of his life, he installed the regular succession of mistresses who ministered to his extra-marital requirements. He had a growing bank account. And, as his choice by Sir Ernest Cassel proved, he was in demand as a negotiator.

It was always Gulbenkian's policy never to let a past insult or injury deflect him from a deal with the person or persons who had inflicted it. So despite the way in which he and his fellow Armenians had been treated by the Turks, he had no hesitation in negotiating with them. He believed in an old Arab saying: 'The hand you dare not bite, kiss it.' He also believed (rightly, as it proved) that the Young Turks' revolution had really changed nothing in Turkish administrative life, and that the skills he had learned at bribery and corruption in the old regime, before the Armenian massacres, would stand him in good stead when he negotiated on behalf of British interests. What the British wanted to do was to set up a bank in Constantinople and infiltrate themselves into an influential position in the Turkish state through ordinary banking activities. The bank was indeed set up and called the National Bank of Turkey, but Gulbenkian immediately set out to convince Sir Ernest Cassel that the last thing they should embark upon was banking, an activity which in Constantinople was so crooked that no upright British financial institution could possibly compete. What they should go out for, instead, were the oil concessions in Mosul and Kirkuk.

Sir Ernest protested that his colleagues were not interested in oil, and that, in any case, there were already other interests in the field. Rear-Admiral Colby Chester, for instance, had struggled back into the running and by his imperturbable admixture of good humour and financial glad-handing, he had persuaded the Ministry of Public Works to restore his lost concessions to him and his syndicate, the Ottoman-American Development Company. Ratification of the concession rights was due to go before the Turkish parliament in 1911. Sir Ernest cited Chester's success as an insuperable obstacle, but Gulbenkian, who had heard in the bazaars that Chester was referring to him as 'an Armenian trickster', shrugged him aside. He could easily be taken care of, he inferred.

On the other hand, it would be necessary to do something about the German interests. Even though the new Turkish government now repudiated their right to build a railroad through to Baghdad, the Germans insisted that the concession was still valid in international law. It would be no use fighting them, Gulbenkian said; far better to compromise and bring them in. So persuasive were his arguments that Cassel and his colleagues in London soon found themselves agreeing to the formation of a company, African and Eastern Concessions (later changed to the Turkish Petroleum Company), in which the £80,000 share capital was divided up as follows: 20,000 shares to the Deutsche Bank (the German interests), 28,000 to Sir Ernest Cassel and the Turkish National Bank, and 32,000 to Calouste Gulbenkian. Within a few months Gulbenkian had passed over 20,000 of his own shares to his friend, Sir Henri Deterding, in the Royal Dutch-Shell organization, keeping 12,000 shares (or 15 per cent of the holding) for himself.

They were all set to go. Coolly ignoring the anxious protests from Admiral Colby Chester, now aware of the intrigues going on around him, Gulbenkian went to work on Turkish officialdom.

By finessing his way through the corridors of Turkish power on the hills of Pera, Gulbenkian had made sure by 1912 that the Turkish Petroleum Company had control over the oil and mineral rights along the thousand mile strip of territory in Iraq previously in dispute between Turkey and the Germans (who were now partners in the Turkish Petroleum Company). So this bone of contention was disposed of. But now he started on the trickiest part of his negotiations, those by which he set out to gain the oil concessions for Kirkuk and Mosul. These were part of the area which had been promised to Admiral Colby Chester and his Ottoman-American Development Company, and the American angrily declared that not only would he fight to retain them but that he had been promised the support of the US government in his battle for the rights to the oil of Mesopotamia. No one took much notice, least of all Calouste Gulbenkian. He had a much more threatening crisis on his hands.

In London the two great proponents of oil-for-the-Navy were

beginning the manoeuvres that would culminate in the British government's purchase of control of the prospering Anglo-Persian Oil Company. Winston Churchill, the first lord of the Admiralty, and Admiral Lord Fisher, the active head of the Royal Navy (as first sea lord), were determined that if they *did* take over control of Anglo-Persian it must become not only the most powerful oil company in Britain but also in the whole of Europe. Therefore in no circumstances must the potentially invaluable oilfields of Mosul and Kirkuk be allowed to fall into any hands but those of Anglo-Persian. The British government had no qualms about Colby Chester, whose claims, no matter how valid, they were convinced would be bypassed. But they had a healthy respect for Calouste Gulbenkian's negotiating genius on behalf of himself and his partners. The order went out: Gulbenkian must be stopped. It was decided that the best method would be to eliminate the Turkish National Bank interests (Sir Ernest Cassel, et al), the Royal Dutch-Shell group (headed by Henri Deterding) and Gulbenkian from the Turkish Petroleum Company, and that Anglo-Persian should take over their shares. Then Anglo-Persian would join forces with the Deutsche Bank to procure the Mosul-Kirkuk concessions between them.

To this end they began with a little high-level arm-twisting. Emissaries came and went between the English directors of the company and the British government, and when they hesitated Churchill played his trump card – snobbery. It was indicated to them that not only were they being unpatriotic but that Buckingham Palace was concerned at their reluctance to cede their shares.

The stratagem worked. Sir Ernest Cassel told Gulbenkian that he and his colleagues had decided to place their 28,000 shares at the disposal of the British government.

The difficult position in which I found myself can easily be imagined [wrote Gulbenkian], the more so as during these pourparlers I received a telegram from Mr Alwayn Parker (a Foreign Office official, then in charge of the Near East Department) by which I was informed it had been decided that I should transfer my shares at once. The style was somewhat peremptory. The position became very confused owing to the National Bank and Sir Ernest Cassel and their associates getting out and leaving us with the Germans, who were astonished at these most unexpected developments.[1]

When Henri Deterding of Royal Dutch-Shell heard of the ultimatum, his rage knew no bounds. Vowing that he would never forgive or forget Gulbenkian's action if he turned his shares over to Anglo-Persian, Deterding stomped off to complain to the Dutch government, and asked them to protest against this attempt to interfere with his legitimate business activities. They promised to back him and urged him to hold on to his shares. Gulbenkian prudently decided to hold on to his too. For the moment, anyway. Eventually a compromise was reached.

It was a stiffer compromise than Gulbenkian, who was barred from the discussions by the British government, had hoped would come out of the talks. The Turkish Petroleum Company was reconstituted, and in the rearrangement of shares Anglo-Persian got 50 per cent, Royal Dutch-Shell 25 per cent, and the German group 25 per cent. As for Gulbenkian, all he got was 2½ per cent given to him 'as a gesture' by Anglo-Persian and Royal Dutch-Shell from their own holdings. These were the shares which would one day cause him to be known as Mr Five Percent.

The disappointed Armenian, convinced that he could have done much better had he been allowed into the conference room, bitterly complained that, considering the work and time and money he had put into the project (which had been his idea in the first place), he had been most shabbily treated. Perhaps. But his five per cent would nevertheless one day make him the richest private individual in the world.

On 28 June 1914, the Turkish Grand Vizier, Said Halim Pasha, wrote letters to the British and German ambassadors officially granting the oil and mineral concessions for Mesopotamia to the Turkish Petroleum Company. At once the United States government protested to Constantinople at the way in which its nationals had been bypassed. But it was too late. The papers had been signed. Admiral Colby Chester put his own concession (worthless once again) back into his brief-case and departed for America in disgust.

Not that the Anglo-Germans got any immediate benefit from their coup, for on 4 August 1914, World War I began and the Turkish Petroleum Company's concession was put into abeyance until 1918.

Long before World War I was over, the British and French

governments between them had decided what was going to be done with the old Arab dependencies of the Turkish Empire. As early as 1916, two years before the defeat of Germany and Turkey, the Sykes-Picot Agreement was signed by the two allies in which their zones of influence within conquered Turkish colonial territory were precisely defined; so precisely, in fact, that there was considerable embarrassment when the British army entered Mosul in 1918 and began immediately to sink wells and build pipelines and railways, only to have it pointed out to them that Mosul came within the French sphere of influence. The difficulty was eventually smoothed out at a meeting in Prime Minister David Lloyd George's apartment in London in 1919, when the French premier, Georges Clemenceau, agreed to hand Mosul over to the British zone, in return, of course, for a quid pro quo. In any case, another clause in the agreement expressed French willingness to respect and sanction any British concessions in their zones which dated from before the war, which should have safeguarded the Turkish Petroleum Company's concession. But with the French it was always best to make sure.

During the war several loose ends had been tied up, and the Turkish Petroleum Company was now indubitably British. The German group's 25 per cent holding in the company had been seized by the Custodian of Enemy Property in London at the outbreak of war, and since the British government controlled Anglo-Persian, which owned a further 47½ per cent of the shares, it now had direct control of 72½ per cent of the company's holdings. In addition, Henri Deterding had become a British citizen in 1915 (he was later knighted for his 'war services'), and brought Royal Dutch-Shell's 22½ per cent with him to Britain. And of course Calouste Gulbenkian, with the remaining 5 per cent, was already British. The British government had no 'foreigners' to worry about as it set about preparing the conquered Arab territories for exploitation.

Or so it thought.

One of the most important of the postwar conferences took place at San Remo, in Italy, in April 1920, at which the British awarded themselves the mandates (in other words, the control) over Palestine and Iraq, and agreed to French assumption of the mandate over Lebanon and Syria. To complaints that it was outside the rights of the two allies to help themselves to mandates in this way, and that only the newly-functioning League of Nations had the power to do so, the British foreign secretary coolly replied that this was a misconception.

It rests with the Powers who have conquered the territories, which it then falls to them to distribute, and it was in these circumstances that the mandate for Palestine and Mesopotamia was conferred upon and accepted by us, and that the mandate for Syria was conferred upon and accepted by France.[2]

The United States government was not invited to the distribution party at San Remo, nor was it informed that during the meeting one of the decisions taken had an important bearing on the future of oil exploitation in Iraq. France had agreed to the transfer of Mosul from French to British control, and at San Remo she got her reward. At Gulbenkian's suggestion, the twenty-five per cent German shareholding in the Turkish Petroleum Company which Britain's Custodian of Enemy Property had seized during the war was passed over to the French. Henceforward, she would have a quarter interest in the oil of Iraq and it would transform France from a petroleum-buying power (mostly from the United States) into a nation with oil resources of her own. France was so delighted with the arrangement that she willingly agreed to build a pipeline across Syria to the Mediterranean for the Mosul oil, and waived duties and harbour charges on its shipment to the markets.

When news of this arrangement leaked out in the United States, great was the indignation of the American oil companies, and a furious campaign began in the newspapers against what one of them described as 'this iniquitous carve-up'. Standard Oil Company of New Jersey reported that one of its geological surveying crews, operating on a previously agreed terrain in Palestine, had been pressured out of the area by the British authorities, and others had been refused permission to proceed to Iraq. It was pointed out to the British government that one of the articles of the Arab mandates guaranteed that 'concessions in the nature of a general monopoly shall not be granted', and the US government, in any case, did not recognize the prewar concession granted to the Turkish Petroleum Company.

It was a time when there was something of a panic in the

United States over oil supplies. The automobile had come of age and all America was buying itself a car (there were by this time more than ten million cars and trucks riding US roads), and the price of crude oil and petrol was rocketing. Experts predicted that US wells were running dry, and that unless some form of fuel rationing were introduced, the nation would have to rely on foreign oil to keep its wheels turning. The situation was not helped by cocky articles by British experts in American magazines forecasting the supersession of the US as the chief oil power by Britain, whose Shell Oil Company had already established itself along American highways. 'Britain will soon be able to do to America what Standard Oil once did to the rest of the world – hold it up to ransom!' wrote one brash commentator in the London *Daily News*.

In such a climate of anxiety and indignation, government, oil companies and public were united in their determination that America was not going to be cheated out of some participation in the hoped for Middle East oil bonanza. An acerbic exchange of notes began between Secretary of State Robert Lansing and the British foreign secretary, Lord Curzon, in which America insisted on a share in the fruits of victory while Britain, though accepting the principle, maintained that it did not apply to rights which had been granted before the war.

What rights? It was Washington's contention that if the British were relying upon the concession of the Turkish Petroleum Company, it was no more valid, in fact it was less valid, than that granted by the Turks to Rear-Admiral Colby Chester, also before the war.

Now it so happened that the unsinkable Admiral Chester had just about this time dug his concession out of the bottom drawer and sailed back to Turkey with it. His subsequent activities were to bedevil and confuse the question of the Mosul oil concession, and for a time it looked as if the US government was going to get behind him and back him to the full. Chester once more persuaded the Turkish parliament to grant him oil and mineral concessions, as well as a strip of land for a railway and a pipeline, all the way from Anatolia to Mosul, and then lobbied untiringly among anti-British and anti-French congressmen and senators to secure support at home. He found an ally in another US navy officer who proved to be as dogged as himself, and even tougher and more antagonistic towards his

French and British allies. This was Admiral Mark L. Bristol who commanded the Eastern Mediterranean naval detachment of the US fleet while he was at sea, but became US high commissioner to Turkey when he went ashore. His British and French colleagues were quite well aware that he preferred the Turkish ex-enemy to his wartime allies, and they winced each time he rose to talk at one of their joint conferences. It was known that he had taken to intercepting and decoding their telegrams to Paris and London, and he was likely to quote their own statements back at them. He believed that the United States should have a mandate over the whole of Turkey *including* the former Turkish colonies, and that all oil and mineral rights should be given to Colby Chester. He was un-abashed in his use of US navy ships to bring American oil technicians and business experts to Turkey, and when the question of the British and French mandates was discussed at the Lausanne Conference he lobbied alongside Colby Chester to get the Mosul enclave carved out of the British mandate over Iraq. Both at Lausanne and subsequently at the League of Nations it was decided otherwise. It was only when both Bristol and Chester appeared to have developed embarrassingly close relations with a prowar party in Turkey (which advocated taking back Mosul from the British and the Arabs by force) that the State Department decided to withdraw support from Chester once and for all.

An additional reason was the fact that the British had by now, under constant prodding by Washington, agreed to make room for the Americans in Iraq. It was Calouste Gulbenkian once more who acted as go-between. Fearful for his five per cent, unswerving in his belief that it was always better to compromise than fight, he used his influence with Sir William Tyrrell at the Foreign Office to persuade the British to accept American shareholders on the Turkish Petroleum Company's board. After due consideration, London suggested that all the American applicants for admission to the Mosul oil concession* should band together under the name of the Near East Development Corporation, and that NEDC should be given a

---

* They consisted of the Standard Oil Company of New Jersey, Standard Oil Company of New York, Gulf Refining Company, Atlantic Refining Company, Pan American Petroleum and Refining Company (Standard of Indiana), Sinclair Oil Company and the Texas Oil Company.

'proportionate' share of the Turkish Petroleum Company's capital.

That did not suit Sir Henri Deterding of Royal Dutch-Shell, who was trying hard amid these tortuous negotiations to sap the predominant strength in the Turkish Petroleum Company of the Anglo-Persian Oil Company and its chief, Sir Charles Greenway, whom he hated. Calouste Gulbenkian's son, Nubar, was now involved in the negotiations and he wrote:

[Deterding] pointed out that on that basis [a proportionate shareout] the Anglo-Persian would have the largest share and could always gang up with one of the other three groups to get a majority. He insisted that all the shareholdings should be equal: one quarter to the Americans, one quarter to the French, one quarter to Royal Dutch-Shell, and one quarter to the Anglo-Persian, all the holdings subject of course to my father's five per cent. This meant cutting down the Anglo-Persian fifty per cent shareholding by half. To persuade them to accept this Deterding suggested that they should receive in lieu a royalty on the oil produced.[3]

This is how it eventually worked out. But it took time. King Faisal of Iraq confirmed the Turkish Petroleum Company's right to drill for oil in his kingdom by granting them a new concession in 1925, and work began at once on finding a suitable field. But it was not until 1927 that a gusher began to spurt oil at Baba Gurgur and set the company on the road to prosperity, and it was not until 1928 that the British, the French and the Americans finally agreed on their shareholdings in what had by then become the Iraq Petroleum Company.*

The British had finally opened the door to the Middle East oilfields and let the Americans put one foot inside. From now on the struggle would be to keep them from bringing in the other.

The full story of the plotting, backbiting, lying and deception which preceded the agreement over the exploitation of Iraq oil is too tortuous to detail here.[4] 'I must say that in the oil business not even one's best friends are to be trusted,' Nubar

---

* The final outcome gave each participant equal shares, as Gulbenkian had suggested. This meant that at board meetings Royal Dutch-Shell, Anglo-Persian, the French and the Americans all had one vote each. (Gulbenkian's five per cent was non-voting.) They still have. Since the American share is now equally divided between Standard of New Jersey and Mobil, this means that the two companies must agree on policy, otherwise their half-vote each nullifies the other. 'The knowledge sometimes brings about a salutary solution to our differences,' a Standard official told the author.

remarks at one point in the negotiations. His father had once said: 'Oilmen are like cats – you can never tell from the sound of them whether they are fighting or making love.' But so far as he was concerned, they were always fighting, and not by Queensberry rules. None of the interested parties (not even the French, who owed their participation to him) hesitated to try to cheat him out of his five per cent, or the proceeds of it. It was one Armenian against the richest corporations, and the smartest lawyers, in the world, and only the bargaining he had learned at the courts of Abdul Hamid enabled him to emerge triumphant.

From the sordid wheeling and dealing of those years one incident involving Gulbenkian needs to be described, for it had important repercussions on subsequent ventures into Middle East oil.

When the original shareholders of the Turkish Petroleum Company had come together in 1914 they had signed a pledge. They had agreed that in all future oil exploitations in the Turkish Empire they would act in concert, and that any fields found by one of the signatories would become the property of them all. When the French had been admitted to TPC in 1920 they had been asked to accept this proviso as the price of membership, and they had done so without a qualm. But when the Americans were told that a similar acceptance was the price of a shareholding, they had jibbed. The US government which believed in the Open Door policy was particularly opposed to such a self-denying ordinance.

In the event, a modus vivendi was found which enabled the American oil companies to accept TPC's restrictive policy and at the same time satisfy Washington (or at least those who did not look at the fine print in the contracts) that the Open Door policy was being observed. However, once oil was struck at Baba Gurgur, and it looked as if there would be other finds in other parts of Iraq, the major shareholders were all for striking out in other directions – but without Gulbenkian.

It was suggested that the other oil groups would endeavour to obtain concessions themselves, leaving me out, [he wrote in his private memoirs] but this was stopped by the threat of legal proceedings to enforce the provisions against competition contained in the Foreign Office agreement of 1914.

Three years of wrangling followed, but finally, at a joint

conference of all the main shareholders in Ostend, Belgium, in July 1928, Gulbenkian's contention was accepted. Not even the Americans could break the self-denying ordinance. The pledge was renewed. None of the shareholders could exploit oil in the territories of the old Turkish Empire without the consent, and the participation, of the others.

There remained one other point to be clarified. What were the boundaries of the old Turkish Empire?

This question could involve untold millions [wrote Gulbenkian's biographer]. The Gulf Corporation of Pennsylvania (one of the American companies [then] involved in the American group of the Turkish Petroleum Company) was already interested in Kuwait and Bahrain, although there had as yet been no oilstrike in either sheikhdom. The British government too had special interests in the Arabian Peninsula, for example treaties with a number of sheikhs who had never admitted Ottoman claims to sovereignty over them.[5]

Which of them was inside the old Turkish Empire and which was beyond its jurisdiction? How were they ever going to define the limits of Ottoman-controlled territory?

No one seemed to be sure. But when the conference looked as if it was going to collapse over the failure to clarify this all-important question, Gulbenkian had an inspiration. He sent out for a large map of the Middle East and laid it on the table in front of the delegates, and then he took a thick red pencil and drew a line round the central area. 'That was the Ottoman Empire which I knew in 1914,' he said. 'And I ought to know. I was born in it, lived in it, served it. If anyone knows better, carry on.'

The delegates hung over the map and peered at the territories cut through by the thick red line. The British were satisfied, because they saw Bahrain, Qatar, Saudi Arabia and the Persian Gulf sheikhdoms were all inside the line. The Americans saw that Kuwait, where they wanted to operate, was outside the line and they were free to go ahead. The others looked and raised no objections.

It was the famous Red Line Agreement which shaped the pattern of oil development in the Middle East for twenty years to come. By agreeing to Gulbenkian's definition, the delegates – though they did not realize it at the time – had ensured the wily Armenian of an income of $50,000,000 a year, for that would be his five per cent share of the oil profits that would be made within the Red Line area in the years to come.

The map with the Red Line was signed and incorporated in the agreement, which the delegates signed before trooping off to a euphoric celebratory lunch.

It was another secret agreement of course, and no one told the Arabs anything about it.

4 · The Wells of Ibn Saud

He was an American citizen who lived in New York and he was proud of his fluent flow of Bowery and Brooklyn slang. But thanks to a childhood spent in the Lebanon, under the shadow of Mount Hermon, he spoke several Arabic dialects equally well. He was a practising Christian but admired the teachings of Mohammed and the tenets of Islam. In Western clothes, clean-shaven save for a trim moustache, he looked like a bright young graduate from one of the fashionable East Coast universities (as indeed he was), but in flowing Arab robes with a short beard added to his chin for desert wear he could easily be, and often was, mistaken for a sheikh. His name was Ameen Rihani and he came to the Arabian Peninsula in 1922 not to look for oil but to see and explore the Middle East and to get to know more about the Arabs with whom he had spent his boyhood. Nevertheless, he did get mixed up with oil, and not very successfully. But his fellow Americans should be grateful to him for his well-meaning intervention. Had it not been for the fact that he backed the wrong man, and persuaded an Arab king to do likewise, the richest oilfields in the world would today be in British rather than American hands.

The Arabian Peninsula, that great, sack-shaped slab of mountain and desert stretching from the Red Sea to the Persian Gulf, and from the Gulf of Akaba to the Indian Ocean, is today three quarters Saudi Arabian territory under the rule of King Faisal, son of the great desert warrior, Abdul Aziz Ibn Saud.* Its capital Riyadh, but the two most cherished jewels in its crown are the holy cities of Islam, Mecca and Medina.

When Ameen Rihani came there in 1922, however, Mecca

* The other quarter is divided among Yemen, South Yemen, the Federation of Arabian Emirates, and Oman.

and Medina and the whole of the Red Sea coast and its hinterland were under the shaky control of a British puppet, King
Husain, who had been dubbed Sherif of Mecca after the Turks
had been driven out of the peninsula in World War I. At that
time, Abdul Aziz Ibn Saud (the future lord of Arabia) and his
bedouin armies had so far conquered only the Persian Gulf
coastal area, the great oasis of Al Hasa, and the desert fortress
of Riyadh, which he had named as his capital and there proclaimed himself Sultan. There would be many bloody battles
with rival claimants to the sherifdom before Ibn Saud and his
wild tribesmen swept westward across mountain and desert
into Mecca and then sent King Husain fleeing into exile from
his Red Sea capital of Jiddah.

In one sense, Ibn Saud was a sort of British puppet himself,
having been given a yearly subsidy of £60,000 during World
War I to fight on the British side against the Turks. Since he
would have fought against the Turks anyway, he had
accepted the bribe and gone on accepting it. So far as he was
concerned, it would make no difference to his plans for the
conquest of Arabia. The British, on the other hand, believed
that with £60,000 they had bought his political loyalty and
safeguarded their plans to keep Husain as Sherif of Mecca and
themselves in control of the peninsula. It was to prove a costly
error.

Why the British ever backed the weak, avaricious and
arrogant Husain in preference to Ibn Saud is one of the
mysteries of Middle East history. In this scorched land of wandering bedouin the man who wielded the sword most mightily
cut and chopped his way to the top, and the man who then
showed mercy to his enemies, but prudently kept them under his
eye, stayed there. Husain was one who paid others to do his
fighting for him, and then got rid of them. No man trusted his
word. No man feared his sword. On the other hand, Ibn Saud had
a sword that was feared and a word that was accepted even by
his bitterest enemies. One day he showed off to Rihani one of his
most valued swords, of chased gold and a silver scabbard. With
this sword, the sultan said, he had dispatched one of his
bitterest foes in a desert battle.

'I struck him first on the leg and disabled him,' he said.
'Quickly after that I struck at the neck – the head fell to one
side – the blood spurted up like a fountain. The third blow at

33

the heart. I saw the heart, which was cut in two, palpitate like that.' He illustrated with a shiver of his hand. 'It was a joyous moment. I kissed the sword.'[1]

This was in the great battle against his hated bedouin rivals, the House of Rashid, but after the fighting was over he took all the remaining members of the Rashid family back to Riyadh, where they lived out the rest of their lives at his expense at court. He married the widow of the head of the house, and adopted his children as his own. And since he was shrewd as well as magnanimous, he bound the successor to the House of Rashid to him by taking his daughter as yet another wife.

By dhow from Bahrain to the tiny fishing port of Oqair, and from there by camel across the desert towards the oasis of Al Hasa, Rihani came for a rendezvous with the mighty warrior. It was a meeting that was to stay in his mind for the rest of his life, for Ibn Saud arrived unexpectedly across the desert by night at the place where the American was encamped (nursing a groin made painfully sore by his camel saddle). He heard the royal caravan approaching from far off across the desert.

Soon the heights on which we were encamped reverberated with the cavalcade of the Sultan. More than 200 camels guggled and growled as they were crouching, while the *ikh, ikh* of the riders and the sound of their bamboos on the necks of their mounts were like the patter of rain in a grove of palms. Soon after the tents were pitched, the fires were lighted, and the tintinabulations of the mortars in the coffee pestles were heard. We hastened forth to meet the great guest, but he was quicker in coming towards us, followed by two of his suite . . . We first met on the sand, under the stars, and in the light of many bonfires that blazed all around. A tall majestic figure in white and brown, overshadowing, overwhelming— that was my first impression . . . And the thing that dominates in him is his magnetic smile.[2]

Ibn Saud took Rihani's hand in his, and it was rapport at first encounter. From that first night meeting came an admiration on the American's part and a trust and respect on that of the desert king which were to have a significant result in the days to come.

Ibn Saud had come down to the Gulf coast from his capital in Riyadh for conferences with Sir Percy Cox, the British high commissioner in Baghdad. The sultan hated the heat and humidity of the coast, and he disliked even more being absent from his wives and concubines, but it was from Sir Percy that he received his annual subsidy, and he needed it for arms and supplies for his bedouin fighters. Therefore he was willing to talk to the high commissioner about the future of Arabia, though he was resolutely determined to make no promises about his own role in the development of it.

When Sir Percy and his party arrived – weeks later than they had indicated – they found that Arab hospitality had gone to great lengths to make them comfortable, though Rihani suspected that they did not appreciate it. Chairs and tables and real camp beds had been found with which to furnish their tents, together with bowls of fresh fruit, and even bottles of Perrier water. There was a great banquet at which a whole camel was roasted. At the dinner given by the Englishmen for the sultan, Sir Percy and his aides appeared in tuxedos. Ibn Saud left early, for he knew that the visitors were anxious to drink brandy and smoke cigars, and that they could not do in his presence.

From the start the discussions went badly. The main subject of the talk (other than the subsidy) was to define the northern borders of Ibn Saud's conquered territory with the new king-dom of Iraq. But on the day after his arrival, Cox sent a note across to the sultan, roughly scribbled in pencil, in which he asked Ibn Saud to give him written assurances of his friendship towards Britain. The terms of the letter the sultan was asked to write were set out in precise detail in an accompanying note. If His Highness returned him the required letter forthwith, Cox added, he 'could then be assured of His Majesty's govern-ment's protection' in the territory he now held.

Ibn Saud was furious. Furious that anyone would dare to send him a scribbled note in pencil, he whose own communica-tions were masterpieces of verbal and calligraphic splendour. Furious that the Englishman should suggest that he was not capable of protecting himself.

'We are only afraid of Allah!' he roared.

It was in this curdled climate that a strange figure appeared on the scene. He waded ashore from a dhow which had brought him, his interpreter and a Somali servant across from Bahrain.* He was a European but he wore over his ordinary clothes 'a

* Bahrain is an island about twenty miles off the Saudi Arabian coast. In those days it was an emirate under British suzerainty.

thin aba [cloak] which concealed nothing, and over his cork helmet a red kerchief and ighal [cord] which made his head seem colossal. But in this attempt to combine good Arab form with comfort and hygiene he certainly looked funny.'[3]

He was a New Zealander named Frank Holmes who had served in World War I in the British navy, but before that had worked as an engineer in the Far East and numbered Herbert Hoover, subsequently president of the US, as a colleague and friend. Now he represented a London syndicate and he was looking for an oil concession which he and his associates intended to prepare for exploitation, and then sell off to the highest bidders preferably the Americans. Rihani, not overly enamoured with the haughty British visitors himself, took immediately to the New Zealander, who shared his growing admiration for the great Ibn Saud. When the sultan sent for Rihani's examination and opinion a clumsily-written Arabic document in which Holmes, through his translator, made his application for an oil concession, the American set to work at once to reword it and at the same time hinted to Ibn Saud that the New Zealander was a man worth dealing with.

When Sir Percy Cox heard the news that there was a stranger in the camp, and one, moreover, seeking an oil concession, he at once called in Holmes to see him, and said: 'Go slow about the concession. The time is not yet ripe for it. The British government cannot afford your company any protection.'

But Holmes was no fool, and he knew that Sir Percy had cabled to London and Baghdad to inform his friends in the government-owned Anglo-Persian Oil Company that an entrepreneur was trespassing on what he considered to be Anglo-Persian's territory.

'Yesterday, Abdul Latif Pasha [Ibn Saud's adviser] showed me a letter,' wrote Rihani in his diary on 30 November 1922, 'which he has just received from his friend Sir Arnold Wilson (president of the Anglo-Persian Oil Company at Abadan*) in which he says he is coming soon to see the sultan and "maybe we can strike a deal about oil".'[4] He confirmed that Wilson had been alerted by Sir Percy Cox.

* He was in fact Resident Director. This is the same Arnold Wilson who, as a young Army lieutenant, had seen the first oilwell gush on the D'Arcy concession in Persia in 1908.

As Rihani remarked, though Cox had told Holmes his application was inopportune, 'evidently it is not untimely for the APO [Anglo-Persian Oil] to negotiate for a concession'.

Not for the first time, a British diplomat had decided that he would squeeze out the national of whom he did not approve in favour of a national of his own class. Major Holmes may have been a British subject, but he was also a brash colonial, whereas Sir Arnold Wilson was an old friend and former colleague. That plus Anglo-Persian's interest in the Saudi concession made him determined to frustrate Holmes's efforts. To the New Zealander's despair and Rihani's fury, the British high commissioner used the future of Ibn Saud's subsidy as a means of pressuring him against Holmes's application.

Sir Percy Cox has asked the Sultan to write a letter to Major Holmes saying that he cannot give his decision till he has made certain inquiries of the British Government and consulted them about the matter [Rihani wrote in his diary on 2 December 1922]. He sent a copy of the letter to be written, with a lead-pencil note to Dr Abdullah [the sultan's secretary] saying: 'Will the Sultan please write letter in the above terms to Major Holmes and send me a copy of it . . .' Sayed Hashem said that three times the Sultan had refused and three times the High Commissioner insisted.[5]

But Ibn Saud wanted his £60,000 and the letter was written.

So it might have ended there, with the defeat of Holmes and the entry of Anglo-Persian into the oilfields of Saudi Arabia. Ameen Rihani, however, was not only angry at the treatment the New Zealander had received, but he was also ashamed of Ibn Saud for knuckling under to the high-handed British diplomat. In fact, so far as the sultan was concerned it did not matter who got the concession so long as he received a rental payment in return – until, that is, Rihani appealed to his sense of pride by writing him a letter on 10 December 1922, in which he said:

Your Highness is a sovereign in your own land, and you have a right to give a concession to whatever company you please, so long as it is English. Your pact with the British Government does not bind you to accept the company they prefer. Here are two English companies, one of them practically owned by the British Government, while the other has nothing to do apparently with politics, is free from all government influence, and you have a right to your own choice in the matter . . . The least of politics with capital the better for Arabia. Concessions on a purely business basis and with a purely business motive, without any political tags to them, or any lead-pencil suggestions from British officials concerning them— these are best for the Arabs and for the English.[6]

The letter hit its target. If Rihani considered Ibn Saud had knuckled under, then the sultan must demonstrate that he had not. But Major Holmes by this time had left the Gulf and installed himself disconsolately in Baghdad. There, some weeks later, Rihani saw him again. The gangling New Zealander told him that he was having tea next day with Lady Cox, the high commissioner's wife, in order to say goodbye to her before he left for England. When Rihani asked him why he did not go back and ask for his concession, Holmes shrugged his shoulders. 'My own government is against me,' he said.

I will give you a letter to [Ibn Saud] and I am certain you will get the concession [Rihani replied]. Never mind what Sir Percy says . . . By all means accept the invitation of Lady Cox to tea and tell her you are going back home . . . Say goodbye too to Sir Percy. For if he suspects you are going back to Al Hasa he might get ahead of you to the Sultan with one or two of those lead-pencil notes . . . Goodbye and good luck.[7]

Major Holmes came back to Ibn Saud's primitive sultanate once more a few weeks later, and the following spring he and his syndicate were granted the oil concession for 60,000 square miles in the desert territory stretching from the shores of the Gulf to the edge of the fertile oasis of Al Hasa. The rental agreed between the syndicate and Ibn Saud was £2,500 a year for the right to survey and drill over what has since become one of the richest oilfields in the world. 'And everyone seemed satisfied with the deal' wrote Ibn Saud's friend, Harry St John Philby, later.

Unfortunately, though the oil was there, Holmes failed to find it. The syndicate did not have much money and its surveys were superficial. Nor could Holmes persuade any of the big American companies (as he had hoped) to take over the concession. From their point of view, Saudi Arabia was too remote, there were no roads or harbours, and fighting was still going on between Ibn Saud and his rivals. After 1927, Holmes stopped paying the annual rental to the sultan and allowed the concession to lapse.

Thus the Holmes concession would be nothing more than a footnote in the story of Saudi Arabian oil were it not for the fact that, failure though it proved to be, it thwarted Anglo-Persian Oil at a moment when that company was ready to take up the concession. Had it done so, almost certainly its experts would have found the oil which had eluded the New Zealander's syndicate. The world's most profitable oilfield would have passed into British hands. The high-handedness of Sir Percy Cox, and Ameen Rihani's antipathy towards him, saved it for the Americans.

5 · 35,000 Gold Sovereigns

By 1930 Abdul Aziz Ibn Saud had killed, captured or put to flight all his enemies, and the land he ruled as King of Saudi Arabia stretched all the way from the Red Sea to the Persian Gulf. Mecca had given up without a fight. Jiddah, the great Red Sea port, had surrendered after the British puppet, Husain, had fled into exile. But there had been bloody battles against rival claimants to lordship of Arabia, and Ibn Saud owed his victory not only to his own strategy and his tactical skill on the battlefield, but also to the fanatic bravery of his Wahhabi bedouin troops and their perfervid spearheads, the *ikhwans*. Like many a Christian general, Ibn Saud had found that a mixture of the sword and the Holy Book worked wonders on the battlefield, and the *ikhwans*, who believed in every word of the Koran, who considered any Muslim who did not an infidel, and had died joyously in battle because they were going straight to paradise, had shed their blood most efficaciously on his behalf.

Now, as their reward, they were allowed to impose their puritanical regime upon the new kingdom, and they did it with a pious passion that was to leave its mark upon Saudi Arabia right up to the present day. Unlike Moses, they did not raise tablets upon which the Commandments of the Prophet were written, for anything suggesting a graven image was henceforward forbidden. But the Commandments were no less stringent because they were spoken. Pictures were torn down from the walls. Statues and monuments were destroyed. *Thou shalt not smoke*. A man caught with a cigarette in his mouth or the smell of tobacco on his clothes was taken into the bazaar square, spreadeagled on the ground by burly slaves, and subjected to forty lashes. *Thou shalt not drink spirituous liquor.* An offender received sixty lashes, or had his hand chopped off at the wrist. *Thou shalt not commit adultery – at least during daylight.* The *ikhwans* and their *ulemas* (priests) believed that man was not ordained to have sexual relations unless darkness had fallen, not even with one of his wives or concubines, and a transgression was subject to the lash. Adultery during Ramadan, the Muslim fast, was punished by a ceremonial execution, and all Jiddah or Riyadh or Al Hufuf was summoned to watch a black slave swordsman lop off the offender's head. (The woman did not merit such a ceremony, and was stoned to death.)

Women were banned from the streets. A man who wore silk robes or walked with a swing of the shoulders was reprimanded for arrogance. It was forbidden to laugh out loud or sing or otherwise make music. And a ceremonial beating awaited anyone who failed to stop whatever he was doing five times a day to hurry away to the statutory prayers at his mosque. Allah be praised, it was a saturnine and sanctimonious regime which had descended upon the Arabian Peninsula.

Not that the new king was affected by it, for his own particular pleasures did not suffer under the round of prohibitions. He had four to which he was particularly attached. True, he was not indulging for the moment in one of them, the joy of fighting hand to hand with an enemy, but that gave him more time for the other three. The first of them was simply talking, gossiping with his courtiers about old enemies he had killed and captured, discussing world politics with his friend and adviser, the Anglo-Irishman, H. St John Philby, or generally reminiscing about the women he had loved and the permutations and techniques of sexual relations. 'Women and world politics continue to divide the honours as prime subjects of the king's conversation at private sessions,' Philby reported in 1930. The second of his pleasures was hunting. For long he had gone out on hawking expeditions into the desert after gazelle, buzzard and game birds, but now the motor car had arrived in Saudi Arabia and it had changed everything. Instead of a hawk to chase and peck out the eyes of the hunted animals and bring them within range of the guns, the automobile now enabled the hunters to chase the quarry and outdistance it. With the help of his new car and a fast-driving

Italian chauffeur, Ibn Saud and his companions managed to thin game in the Arabian desert down to extinction levels. The ostriches which roamed the foothills until 1931 were then wiped out, and the last of rare breeds of oryx and gazelle were driven into the mountains. Philby once saw the king wipe out a herd of rare gazelles in fifteen minutes, and there was no doubt of the great joy he got from the processes of extermination.*

But the third and undoubtedly the most savoured of his pleasures was sex, and since he agreed with the *ulemas* that it should never be indulged in during daylight hours, when he considered it indecent, he encountered no disapproval from them of his sexual activities with the opposite sex, marathon though they were. From the age of eleven until a few weeks before his death in 1953 at the age of seventy-two he had sexual relations with a different woman every night, save during battles and while the *haj* or pilgrimage to Mecca was in progress, when all good Muslims are pledged to eschew sex. Each night at nine-thirty he would glance at his watch, say adieu to his guests, and depart for his harem. He had married for the first time at the age of fourteen, and it is likely that he loved his young bride for he grieved deeply when she died in an epidemic. But thereafter he gave himself over to what Philby called 'his marked tendency to uxoriousness'. He is said to have told the Ruler of Kuwait that the pleasure he found 'most worth living for' was to 'put his lips on the woman's lips, his body on her body and his feet on her feet'. There was some comment, after this remark became known, that this, in his case, if taken literally, would have severely limited his choice of females in Arabia, since he was six feet six inches in height. That all he asked for from a woman was sexual relations (and children, of course) was indicated by his remark that he had never had other than small talk with a member of the female sex, and until American oilmen in Arabia began to bring their wives, and were sometimes entertained by the king, he had never seen a woman eat or drink, not even his mother.

The king never exceeded the quota of four wives allowed him by Koranic law, but sometimes he found it politic to marry the widow of a defeated chief or the daughter of a tribal leader whose allegiance he was seeking, so the quadrumvirate of

* Philby rescued one gazelle head and sent it to the British Museum.

wives remained constant but their identities changed frequently as he solemnly pronounced one divorced before wedding the other. He had sexual relations with all of them, though he confessed that with some he never even bothered to take off their veils.

In the months following his conquest of Arabia, he was in his sexual prime, and though it seemed to make no overdue demands on his physical reserves, it certainly depleted his treasury.

Each of his four wives having a house of her own with a full complement of servants, slaves and attendants, in which they received his visits in rotation [wrote Philby], he also had a house of his own . . . which was run for him by four favourite concubines, enjoying a status indistinguishable from that of wedded wives. And in addition he had four favoured slave girls to complete his matrimonial team of a dozen, to say nothing of his right to pick and choose from other numerous damsels at his disposal.[1]

Ibn Saud once confessed to Philby (in 1930) that he had married 135 virgins, to say nothing of 'about a hundred others' up to that time. But he was finding them a desperate drain on his financial resources, especially as he also had to keep their servants and any children they produced. He had resolved to limit himself to two new wives a year from now on (since it was wives who were most expensive) but even this would not solve the monetary troubles in which he and the new kingdom found themselves. Desperately he asked Philby what he should do.

Harry St John Philby was an Arabist scholar and explorer who had once worked for Sir Percy Cox, the British high commissioner in Baghdad, but had resigned from the Colonial Service, disenchanted by Anglo-French treatment of the Arabs. He had won the friendship of the desert king and become the only man at court who dared to argue with him, or proffer advice unasked and have it taken. He had turned Muslim and built himself homes at Mecca and Jiddah. To help keep his wife and children in England, he had also secured for himself the lucrative Ford car agency for Saudi Arabia and sent Model As out on to the desert tracks where before only human and animal feet had touched.

Philby knew that no economy in wives would save the kingdom from bankruptcy, and that, in any case, the royal resolution to abstain would be abandoned the moment a particularly pleasing concubine or slave began to plead for marriage. Saudi Arabia's financial problems were much more serious than that. The main income of the country was garnered from the pilgrims who flocked each year to Mecca, and because of the unrest in Arabia and the depression abroad only small numbers had come in 1930 and even fewer were expected in 1931.

One day in the autumn of 1930 Philby accompanied the king on a desert drive, and soon realized that Ibn Saud was worried. He asked what was the matter. The king replied that the financial situation was so bad that the government could see no way to make ends meet, and was at the same time confronted by difficulties that it could not hope to cope with.

I replied, as cheerfully as possible in the circumstances [wrote Philby later], that he and his Government were like folk asleep on the site of buried treasure, but too lazy or too frightened to dig in search of it. Challenged to make my meaning clear, I said I had no doubt whatever that his enormous country contained rich mineral resources, though they were of little use to him or to anybody else in the bowels of the earth. Their existence could only be proved by expert prospection, while their ultimate exploitation for the benefit of the country necessarily involved the cooperation of foreign technicians and capital. Yet the Government seemed to have set its face against the development of its potential wealth by foreign agencies.[2]

Instead of moaning about poverty, Philby went on, Ibn Saud should think about the Koranic saying: 'God changeth not that which is in people unless they change that which is in themselves.'

'Oh, Philby,' exclaimed the king, 'if anyone would offer me a million pounds, I would give him all the concessions he wanted.'

'Well,' replied Philby, 'it won't be as bad as that. But nobody will give you a million pounds without a preliminary investigation of the potential resources of your country. Yet there is a man in Egypt now – an American whom you did not take the trouble to meet when he visited Jiddah a few years ago. He has recently visited the Yemen and done a good deal to help that country in various ways. He could help you too, for he is a very rich man with important contacts in the American industrial community. I suggest you should meet him as soon as possible.'

The man to whom he was referring was a US philanthropist named Charles Crane, whose passionate love affair with the Arab people had survived snubs and setbacks and even armed attacks (during one of which a companion travelling with him through the desert was shot dead). His affection for its people left him surprisingly ignorant of some things about Arabia, however. He arrived in Jiddah the following February and had several discussions with the king and his ministers. Just before departing Ibn Saud made him a gift of two magnificent Arab stallions. Crane's return gift was, of all things in a land whose oases grew nothing else, a box of Californian dates. But he did give useful advice about the exploitation of mineral resources and, within six weeks, had dispatched to Jiddah a Vermont mining engineer named Karl Twitchell, who, at Crane's expense, would prospect the desert for oil possibilities.

Karl Twitchell, in the words of the official historian of the Arabian American Oil Company, 'went up and down Arabia sniffing for oil seeps and quartz outcrops' and his reports in 1932 'so encouraged Ibn Saud about oil and gold prospects that he commissioned Twitchell to communicate to oil or mining companies in the United States, Arabia's willingness to discuss concessions.'[3]

What neither Twitchell nor King Ibn Saud knew was that Standard Oil of California officials had been trying to contact the king for months in the hopes of talking concessions. But they had been stalled and hindered in their efforts by none other than the dashing New Zealander, Major Frank Holmes, whose image during these events becomes rather less burnished than it was when Ameen Rihani had recommended him to Ibn Saud in 1922.

Having failed to find a buyer for the Al Hasa concession Ibn Saud had granted him, and lacking the financial resources to explore the terrain adequately himself, Holmes had simply let the concession lapse, without informing the king or paying the modest rental of £2,500 which was due on it. All communications from Arabia were unanswered, and it was assumed that Holmes had gone back either to England or New Zealand. He was, in fact, only twenty miles away, on the island of Bahrain, just off the Arabian coast. There in 1925 he was granted an oil concession by the ruling sheikh, with the approval of the British resident (for Bahrain, like most other sheikhdoms and emirates in the Gulf, was under British control), and the concession was renewed in 1927. In that year Holmes sold an option on it to the American company, Eastern Gulf Oil, a subsidiary of the Gulf Oil Company. The US oilmen sent a geologist, Ralph Rhoades, to look the territory over and they were delighted. Rhoades reported that Bahrain had a perfect geological structure, and recommended that drilling for oil should begin immediately.

But there was a snag. Bahrain was within the red line drawn on the map by Calouste Gulbenkian in his definition of the confines of the old Turkish Empire. The Red Line Agreement stipulated that no members of the Iraq Petroleum Company could exploit a concession within the area independently of the others. Gulf at that time was one of the consortium of American companies which had a share in IPC. So Gulf offered the Bahrain concession to IPC, and promptly IPC turned it down on the grounds that *their* geologists considered the oil prospects poor.

It was the first serious indication to the Americans that Gulbenkian's red line was going to be a halter round their necks, and Gulf shortly afterwards bowed out of IPC in order to regain independence of action, as did several other US companies. But it was too late to save Bahrain. Gulf sold out their option to Standard Oil of California (Socal), who almost immediately ran into trouble with it themselves.

No one had told Socal that in the British-controlled areas of the Middle East only British firms were allowed to operate, and it took them two years to arrange a way around the regulations. A company had first to be registered in Canada, then an office established in Britain run by a British subject. It was stipulated that an active director of the company must not only be British but officially approved by the government. The majority of the employees must be either British or Bahraini, and the company undertook to appoint in Bahrain itself a chief local representative, vetted and approved by the British government, who would never, but never, approach the Ruler of Bahrain except through the offices of the British Resident.

It was not until 1 August 1930, that Socal formally assigned its concession to its Canadian subsidiary, Bahrain Petroleum

Company (Bapco), and at the same time it announced the appointment as its chief local representative of Major Frank Holmes.

Bahrain's first well came in, confounding IPC's geologists, on 14 October 1931, but long before that Socal's visiting oilmen knew that there was oil there. But more important than that, their soundings told them that if there was an oil pool under Bahrain there was probably an even bigger pool under the terrain they could see vaguely shimmering in the heat, twenty miles away – in Saudi Arabia.

A Socal geologist named Fred A. Davies, a huge full-back figure of a man from Pennsylvania, was so convinced that Saudi Arabia was where most of the oil was waiting for them that he galvanized home office in San Francisco into agreeing to go after the concession. It was decided to approach Ibn Saud, and the man who could take them to him, or so they believed, was obviously Major Frank Holmes, who was now on their payroll. Holmes was told to set up a meeting, and he agreed at once to do so. What he does not seem to have told them is that he had failed to pay the rental on his own lapsed Al Hasa concession, and was therefore not exactly anxious to meet his creditors. While Socal officials waited in the steamy heat of Bahrain, twenty tantalizing miles away from the potentials of Al Hasa, Holmes fobbed them off with excuses. First he told them that King Ibn Saud was fighting a campaign, then that he was mourning a favourite wife. Eventually, tired of waiting for a rendezvous that looked more and more like a mirage, Davies and his fellow Socal officials sailed for home.

In the meantime, the company's adviser on foreign relations, Francis B. Loomis (who had once been an under-secretary of state under Theodore Roosevelt), had heard of Harry St John Philby's influence with the Saudi monarch and had written him a letter asking him to arrange a meeting for Socal representatives with the king. A few days afterwards, at a luncheon party in Washington, Loomis found himself talking to Karl Twitchell, back in the US for a vacation. He mentioned Socal's interest in a Saudi concession.

'Then why haven't you been in touch with the king,' asked Twitchell. 'Yours are just the people he wants to meet.'

Loomis wasted no time telling him about the procrastinations of Holmes,* but at once asked Twitchell to return to Saudi Arabia with a Socal negotiator and act as go-between in the discussions with the king. Socal chose as its negotiator a plump, forty-year-old lawyer and land-lease expert from California named Lloyd N. Hamilton. The two men arrived in Jiddah aboard the S.S. *Burmah* on 20 February 1933, and they brought their wives, Airy Hamilton and Nona Twitchell, along with them. Hamilton had the slightly glazed look of an innocent American abroad, but appearances were deceptive, as his Arab opposite numbers and rival oil companies would discover. It was a fateful arrival. 'Change,' as Aramco's official historian puts it, 'entered in a suit of wrinkled whites and a sun helmet'.[4]

Meanwhile, urgent conferences were being held in London over the whole question of oil concessions in the Persian Gulf. To say that the Anglo-Persian Oil Company was concerned over US incursions into an area hitherto considered a British bailiwick is to put it far too politely. The new chairman of the company, Sir John Cadman, was sternly demanding of his British associates in the Iraq Petroleum Company why they had let the Bahrain concession go to the Americans, when they had the prior right to take it up themselves. It was explained to him that IPC geologists had not believed it possible for oil to be found in porous limestone of Cretaceous age (the deposits in Iraq and Persia being in much older rock formations). He is said to have replied: 'In that case, I suggest you sack our geologists and hire Americans.'

With the oil of Bahrain, it would have been easy. The ruler was under British control. All IPC had to do, under the Red Line Agreement, was take over. But with Saudi Arabia the situation would be much more difficult. The time was past when King Ibn Saud would be bullied by the British. He would sell to the highest bidder. The Saudi concession would now have to be fought for with an American rival already successfully established just across the water in Bahrain, and one moreover which had signed no prior agreements with anyone and was not bound by the Red Line or any other agreements. Cadman

---

* Holmes was only too right in his hesitations, so far as he was personally concerned. He arrived in Jiddah during subsequent Socal negotiations and was at once told to pay up his concession rentals or get out within forty-eight hours. He got out.

frigidly indicated to his subordinates that IPC would not be dilatory or lacking in combativeness on this occasion. From his point of view, the situation was doubly dangerous—commercially and politically. If Socal succeeded in Saudi Arabia, it would not only bring an American oil rival into an area where British petroleum interests had so far been dominant, but, since the flag followed trade, it would also bring US influence into the Persian Gulf, a zone which Britain considered vital to her imperial interests and was determined to keep walled-off from any foreign interference.

If the Anglo-Persian chief was aware of the important strategical as well as tactical advantages of securing the Saudi concession for Britain, however, no one in IPC seems to have explained them to the British negotiator the company sent to Jiddah, Stephen Hemsley Longrigg. Longrigg is a most able and experienced veteran of the Middle East oil industry, and he has written a most authoritative history of how it evolved.[5] But in that account he devotes only one paragraph to his own and IPC's part in the negotiations for Saudi Arabia's oil. His reticence is understandable for the company blundered badly. All he says is:

> The Iraq Petroleum Company had this time decided to contest the issue. Their representative (the present writer) arrived at Jiddah to find negotiations in progress between Hamilton [of Socal] and the Saudi ministers and was invited to make his own offers. Both negotiators interviewed the King, both advanced their proposals, each was assured that his company and his nationality would, all things being equal, be more acceptable to the Saudi King. But the IPC directors were slow and cautious in their offers and would speak only of rupees when gold was demanded. Their negotiator, so handicapped, could do little; and agreement was reached without difficulty between Hamilton and Shaikh Abdullah Sulaiman on 29 May 1933 for a 60-year concession for the Al Hasa province.

But there was much more to it than that.

IPC might have sent its negotiator to Jiddah even later in the day had not Harry St John Philby sent a message to Anglo-Persian in London telling them that Socal representatives were about to arrive to make a deal. This was just after he had received a note from King Ibn Saud (dated 23 February 1933) saying, after a friendly preamble:

I am confident that you will protect our interests, both economic and political, just as you would protect your own personal interests. So I shall expect your assistance in this matter and also I shall expect you to give me the benefit of your personal advice, which will be treated as confidential and with all consideration.[6]

It was to secure a better bargain for the king that Philby urged Anglo-Persian to prod IPC into sending a negotiator, on the correct assumption that competition for the concession would force up the price that was offered for it. But by the time Longrigg arrived, the situation had radically changed. In his conversations with Philby, Lloyd Hamilton, the Socal representative, had quickly divined that the Englishman, like Ibn Saud himself, though on a humbler scale, had a money problem. The Ford Agency was not bringing in as much profit as he had hoped, and there were fees to be paid for his son, at Cambridge University, and his three daughters at school.* Hamilton therefore made a discreet proposition. If Philby would help Socal to secure the concession, by pressing the American case with the king and his ministers, and by keeping Hamilton informed of any rival offers, he would be paid a fee of £1,000 a month for a minimum of six months, added to which there would be bonuses for him when the concession was signed and when oil was struck.

Philby accepted it without, it appears, any hesitation. The only stipulation he made was that no one should be told of the arrangement. Hamilton did not even tell his own wife, or Twitchell. Philby did not inform the king, or his minister, Abdullah Sulaiman. Longrigg was so completely in the dark about what was going on that when he arrived he asked Philby if he would like to take over the negotiations for IPC. Philby refused on the grounds that it would affect his 'impartiality'.

Despite the fact that he was now on the Americans' payroll, Philby might have kept the IPC representative in the game had not Longrigg decided at one of their meetings to take his fellow countryman into his confidence. The two men had once worked together in the Colonial service, and they had sat on the same side of the table at many an Arabian conference; Longrigg had no reason not to trust his compatriot. He even appears to have

* The son, Kim, grew up to become a departmental chief of British Intelligence in World War II; in 1963 he defected to the Soviet Union, where he is now an officer of the KGB.

felt that Philby, for all his declarations of 'impartiality', would do his best for him and for Britain in the negotiations. Now Longrigg confessed that IPC was not really interested in Arabian oil at all. The company did not care whether any was found or not. All it was anxious to do was prevent the concession from falling into the hands of the Americans. Petroleum had slumped to ten cents a barrel in the world markets and there was a trade depression everywhere, and IPC had all the oil it could handle from its wells in Iraq.

This was jolting news for Philby. For Ibn Saud's sake, he wanted to see Arabia's oil potential exploited as urgently as possible. It reinforced the suspicion he had always had that IPC were not the right people for the concession.

What confirmed it was Longrigg's next revelation. All he had been authorized to offer for a concession was £5,000. Philby told him that he had better go away at once and get the stake raised, and make sure that whatever sum he came back with was payable in gold. Longrigg hurried off to the British Legation to have a cipher telegram dispatched to London, and reappeared subsequently with the news that he could go up to £6,000 but that there would be no question of payment being made in gold. Britain had gone off the gold standard and the Bank of England would never allow it, for it would be tantamount to suggesting that the pound sterling was unstable. The king would have to accept sterling, or the equivalent in sterling-backed Indian rupees.

Longrigg stayed around in Jiddah for some time after that, but Philby eliminated him from the negotiations. So far as he was concerned, it was now a straight contest between Abdullah Sulaiman, the Saudi finance minister, and Lloyd Hamilton of Socal, with Philby acting as the go-between. Years later, the Saudi minister told Philby that 'if he had known at the time that I was actually supporting the negotiations of the California Standard Company for the oil concession in 1953, he would have done his best to block the transaction.'[7] In fact, he was doing Philby an injustice, for the moment he realized that IPC was not a contender worth serious consideration for the concession, Philby did his best, while strongly supporting Socal's case, to get a really generous sum of money out of the Americans. But he was no oilman, and he did not realize the huge potentials involved. He ferried himself back and forth in his brand new Model A Ford between the Grand Hotel, where the Lloyd Hamiltons were installed in a huge room with a brass bedstead and an airless bathroom with cockroaches but without running water, and the king's palace beyond Jiddah's city walls. Hamilton, for all his shrewdness, had expected the negotiations to take only a week or two, but there were endless rounds of conversations and innumerable cups of coffee consumed during which no word about oil was spoken. Then, at Philby's suggestion, Socal was asked for an immediate payment of £100,000, and he was considerably shaken when Hamilton told him that all Socal was prepared to offer was $50,000 against future royalties. Weeks went by, and it was not until 20 April, when the humid heat of Jiddah was boiling up to the temperature of a Turkish bath, that Hamilton met Abdullah Sulaiman and presented his final offer of an immediate loan in gold of £30,000, plus a further loan of £20,000 gold eighteen months later, and a yearly rent of £5,000 gold. Philby knew from his conversations with Hamilton that this was indeed the final offer, and had conveyed that fact to Abdullah Sulaiman, so the Saudis were therefore all ready to accept it when a Reuter's telegram reached them. As the historian of Aramco remarks, it 'threw the whole tangled negotiations into a snarl again'. For 20 April 1933 was the day on which the United States announced an embargo on gold.

Within a few days Lombardi [a Socal vice-president] was cabling from London to hold everything, and the Netherlands Bank in Jiddah had raised the value of the gold pound from $4.87 to $5.60. Within a few days it climbed to above $6—carrying with it the whole price of Hamilton's carefully negotiated agreements.[8]

New negotiations began, as Hamilton tried to turn the Saudis away from gold pounds to paper dollars. They were adamant. It was not until 25 May 1933, that Abdullah Sulaiman came into the king's presence and began reading the long and complicated agreement which he had made with Lloyd Hamilton. It took two days to read, and Ibn Saud, puzzled by its convolutions, went to sleep during most of it. But towards the end he woke up with a start, fixed Philby, sitting on the ground near him, with his one good eye* and asked him what

* Like many another Arab, he suffered from trachoma.

he thought of the agreement. Philby assured him that it was the best Saudi Arabia could have obtained.

'Very well,' said the king, to the finance minister, 'put your trust in God and sign.'

On 29 May 1933, Abdullah Sulaiman for the Saudis and Lloyd Hamilton for Socal signed the concession agreement that was to bring American capital, brains and technical skill into the Persian Gulf and change the course of Arabian history. The door of the Middle East had opened wide enough for the United States, with one foot planted in Bahrain, to get a second on to the sands of the mainland. It did not happen immediately, but it was the beginning of the end of British control of the Persian Gulf. Britain had not only let the Americans into a cherished sphere of British influence, but she had also let slip the biggest Middle Eastern oilfield of them all.

For his part in the negotiations, Harry St John Philby received the warm thanks of King Ibn Saud and the gift of a house, known as the Green Palace, with a garden, not far from the king's palace, beyond the walls of old Jiddah. This enabled him to sell his old house in the city, the Bait Bagdadi, to Lloyd Hamilton as a temporary headquarters for the company's Saudi operations. He had done well by Socal and he had earned his fees and bonuses. Everyone seemed satisfied, and it was only later, when the gushers began to come in, that the Saudis began to claim that they had been cheated.

In fact it was only by a stroke of luck that they received their first payment at all. Hamilton had pledged that the initial payment in gold would be made to the Saudi finance minister within three months of the signing of the agreement. Under the new currency regulations, however, all gold exports from the United States had to be authorized by the Treasury Department. Socal at once made application for a permit to export $170,327.50 in gold to Saudi Arabia. Weeks passed and there was no word from Washington. Hamilton, who had been on vacation in France and Italy, arrived in London in July to learn of the ominous silence, and decided that drastic measures must be taken.

On 26 July, he had Socal officials go into the London gold market and buy 35,000 gold sovereigns from the Guaranty Trust Company. It was an illegal purchase for an American company under the new US regulations, but the future of the concession was at stake. He was more than ever convinced that he was right to have taken the risk when a reply was at last received, on 28 July, from the Treasury Department, signed by the undersecretary, a certain Dean Acheson, formally turning down Socal's application.

The 35,000 black-market sovereigns were boxed and put aboard a P & O liner which sailed for the Persian Gulf on 4 August. On 25 August, three days short of the deadline, Karl Twitchell sat at a table in the Jiddah branch of the Netherlands Bank and counted out 35,000 sovereigns one by one under the wary eyes of Abdullah Sulaiman. Then he got a receipt, and the Americans were in business in Saudi Arabia.

6 · And Then There Was Kuwait

Two hundred and fifty miles north of Bahrain and Al Hasa, at the neck of the Persian Gulf, a wedge of coastline and desert, shaped like the bisected half of a six-pointed star, has been inserted into the frontiers between Saudi Arabia and Iraq. This is the Sheikhdom of Kuwait. It is a car-clogged land of millionaires, of sky-scrapered hotels and ornate government buildings today, and it sits on a vast (if not inexhaustible) pool of high-quality oil. But in the 1930s it was a sleepy pearl-fishing village. As to oil, all that the geologists knew was that there were bitumen seepages at Burgan, in the desert away from the sea, and that underneath viable quantities of oil probably did indeed exist, but no one could say for certain until they had been drilled for.

Into Kuwait in 1930 came a character who has figured several times before in this story, none other than the bouncy New Zealander, Major Frank Holmes, once more in search of an oil concession. He already had a customer for it. When he had sold out his Bahrain concession to Eastern Gulf Oil Company he had agreed, as part of the contract, to obtain a concession in Kuwait should Gulf ever ask for it. In the intervening period, Gulf had been forced to sell off Bahrain to Standard Oil of California because of their commitment to the Red Line Agreement (and their angry frustration at this forced sale was doubly compounded when Socal struck oil in Bahrain in May 1931). They told Holmes they were exercising their option under their contract with him, and they wanted the Kuwait concession. They wanted Kuwait not only because they were confident that there was oil there, but because they believed that in Kuwait they would have a free hand.

Gulbenkian, when he had drawn the red line defining the boundaries of the old Turkish Empire, had left Kuwait on the free side of the line.

This time Holmes needed quick action, for he was short of money, and he arrived in the sheikhdom bearing gifts. For his first serious talk with the ruler of Kuwait, Sheikh Ahmad al Jabir as-Sabah, he arrived in a brand new Sunbeam motorcar, driven by his associate, T. E. Ward, and when the sheikh came out to admire its sleek English lines the New Zealander said, with a flourish: 'It is yours, Excellency.' Slipping the knot of the green gauze veil which he wore to protect his face from the sand and doffing his white cork sun helmet, he followed the ruler into the palace. And all seemed to be well.

'It is interesting to observe the greedy attitude the Sheikh of Kuwait exhibited when I hinted that the company I was representing had an American tang to it,' he remarked to Ward afterwards.

In the next months the greed was fed by gifts which Ward afterwards enumerated in the account he drew up for William J. Wallace of the Gulf Corporation. Four members of the state council received 4,000 rupees, 45,000 rupees were reserved for certain high officials, and 20,000 rupees went into the ruler's private purse.

The concession was as good as granted. Or so Sheikh Ahmad indicated. But soon rumours began to circulate around the bazaars in Kuwait that the British political agent, Colonel Harold Dickson, had been several times to see the ruler, and the sheikh had been reminded of certain 'obligations'. There were suddenly no more conversations between the sheikh and the New Zealander, and further baksheesh was politely turned away.

Gulf were growing increasingly impatient at the sudden halt in the negotiations, especially now that they were aware of the oil strike in Bahrain. But it was not until well into 1931 that they discovered the reason for the delay. As they had done in the case of Bahrain, the British government let it be known that they were invoking an old treaty with the sheikhdom to prevent an American, or any other foreign company operating in the territory; and this time, they inferred, the difficulty would not be overcome by the registration of a US company in

Canada to give a false coat of Britishness to its colours. It would have to be a British company or nothing.*

At which point a genuine British company appeared on the scene right on cue, in the person of A. H. T. Chisholm, representing the Anglo-Persian Oil Company. Just as Gulf was not restricted by the Red Line Agreement in Kuwait, neither was its partner in the Iraq Petroleum Company, Anglo-Persian; and APO had the British government and a sheikhly treaty behind it in the campaign which began to oust Gulf. Chisholm later admitted his company was not primarily interested in finding oil in Kuwait (it should be mentioned again that these were the days of the great world depression, and petroleum sales were drastically down).

But we *had* to get the concession so as to protect our huge investment in Persia [he said]. For Kuwait is nearer [world] markets and, if competitors found oil there, the Persian business would have been undercut and ruined.[1]

He set to work to stop them.

The wife of the political agent in Kuwait, Mrs Violet Dickson, paints a cosy picture of two friendly rivals dabbling in oil concessions in between the social round, while her husband sat with Sheikh Ahmad at the discussions and gave him impartial advice. 'Chisholm in particular,' she writes, 'with his youthful charm, his gaiety and his skill at tennis, was an asset to all our social gatherings.' And Holmes was 'forceful and often rude, but could be good company nevertheless.' And, of course, 'Sheikh Ahmad had grown to trust [Harold] as a friend.'[2]

The situation was somewhat tougher than that. The two men were complete contrasts in types. Chisholm was thirty years old, Anglo-Irish, the son of a former editor of *The Times*, with an honours degree from Oxford, and a year on the staff of the *Wall Street Journal*. He had a family connection with Sir

* To reinforce their argument that Kuwait's oil was reserved for Britain, the government reproduced a letter written by the then Ruler to the British resident in Bushire in October 1913, in which, in return for the promise of British protection from the Turks, he wrote: 'We are agreeable to everything which you regard as advantageous and if the Admiral honours our country [with a visit] we will associate with him one of our sons to be in his service, to show the place of bitumen in Burgan and elsewhere, and if in their view there seems hope of obtaining oil therefrom we shall never give a concession in this matter to anyone except a person appointed from the British Government.' (*Treaties, Engagements and Sanads relating to India and Neighbouring Countries*, Delhi, 1933).

John Cadman, the Anglo-Persian chief, dressed like a character out of a Noel Coward play, sported a monocle, and larded his conversation with the latest Mayfair slang. But behind the foppish front was a cold, calculating mind that would make him (as the author came to know) one of the most successful military intelligence officers in the Middle East in World War II.

Frank Holmes was over fifty and when he walked through Kuwait in his ridiculous solar topee and green face veil small boys would walk behind him, laughing at the white sunshade he carried, and singing:

'Ingaressi bu taila,
Assa yi mutun hal laila!'
(Englishman with hat on your head,
May you die tonight in bed.)

But here again the eccentric outer garments and florid front were a cover for a rough man who knew exactly what he wanted, and was determined to get it. So far as Holmes was concerned, the Kuwait oil concession was a matter of life or death. All Chisholm would get if he failed was a slap on the wrist from his relative on APO's board, but if Holmes failed he knew Gulf would sue and squeeze him dry.

'Holmes and I were at daggers drawn for a year,' Chisholm said later. 'Month after month we lived as neighbours, meeting with invariable courtesy and no rancour, each making calls on the sheikh, each in turn suggesting some new clause and knowing that the other would act accordingly. Every Sunday we met in church.'[3]

By this time the ruler of Kuwait, Sheikh Ahmad, had begun to like the situation. So long as Holmes and his Americans remained in the competition, he had everything to gain. Even if British pressure did force him finally, as he suspected it would, to favour the cool bemonocled Anglo-Irishman over his New Zealand rival the preliminary struggles would drive the price up; and since Sheikh Ahmad had expensive and exotic tastes, especially in women, he needed all the money he could get.

Not that Holmes considered that he and his friends were beaten yet, despite the British declaration. Nor did Gulf. They knew they had a high card in their hand, and on 27 November

1931 they played it. On that date they formally called the attention of the State Department 'to the fact that the so-called British Colonial Office was insisting on the so-called "nationality clause" in the Kuwait concession. This clause in effect prevented anyone except a British subject or firm from obtaining a concession in Kuwait.'[4]

The State Department's reaction was prompt. On 3 December 1931 'the instructions were sent to our embassy in London to make representations to securing equal treatment for American firms.' As a State Department memorandum later pointed out:

These negotiations were long, and complicated at a later date by the fact that the British-controlled Anglo-Persian, which had previously expressed its disinterest in Kuwait, suddenly endeavoured to secure a concession from the sheikh of Kuwait. Here again the Department insisted on the 'open door' policy, and our Embassy in London was assiduous in its endeavour to expedite a settlement, and continuously and frequently pressured the British authorities for action.[5]

Frank Holmes and his sponsors knew only too well why the US ambassador in London was so 'assiduous' on their behalf. Andrew Mellon, the Pennsylvania millionaire, was American envoy to London at the time, and until his appointment he had been an active director of Gulf Oil, and the company was controlled by the Mellon family interests. But though he 'continuously and frequently pressured' the British government to let Gulf take over Kuwait, so long as Anglo-Persian insisted on the concession Whitehall resisted all US demands. And for a time Anglo-Persian was inflexible; they even brought rigging crews over to Kuwait from Persia, ready to start work. As the competition dragged on, Sheikh Ahmad delightedly rejected each company's bid in turn and upped his demands; for more money down, for more rental, and (worst of all in the eyes of both rivals) for a Kuwaiti director on the board of the successful company.

Then one Sunday morning in 1932, halfway through the service in the stuffily hot little Anglican church, his sensitive musical ear cringing under the blast of Frank Holmes's foghorn baritone thundering out the hymns behind him, Archie Chisholm thought of Solomon. Why not do what the biblical king suggested, and cut the baby in two?

About the same time, T. E. Ward, Frank Holmes's legal adviser and a fellow director of his syndicate, had the same idea. The time was right for compromise. He had just read a speech made by Sir John Cadman of Anglo-Persian to a convention at the American Petroleum Institute in Houston, Texas, in which Cadman complained of a 'glut' of petroleum and cut-throat competition among its producers.

Consumption everywhere has decreased [Sir John said], competition is too keen, and prices contain no margin for gain. One possible step towards rehabilitation of trade is evidently the readjustment of supply and demand and the prevention of excessive competition by allotting to each country a quota which it will undertake not to exceed.

It was the first public admission that the oil companies were moving towards the establishment of an international cartel to control petroleum supplies and prices. To T. E. Ward the words used by Cadman 'were not only a keynote for the convention but also for the Anglo-Persian and Gulf Company directors on the subject of equal partnership in Kuwait.'[6]

So almost simultaneously Chisholm and Holmes arrived at the conclusion that it was useless to go on fighting for a monopoly of Kuwait's oil. There is no record that they ever discussed the question of a compromise together, even though, like the colonel's lady and Judy O'Grady, they were petroliferous sisters under the skin. But both of them went to work on their home offices.

Suddenly it was all over. Sheikh Ahmad discovered that not only were the two negotiators no longer bidding against each other, they were actually demanding that he scale down his demands. First they forced him to reduce the royalty on the oil they hoped to extract. Then they rejected outright his demand that a Kuwaiti join the board of directors of the new company. As the price dropped at each meeting, it became obvious to Sheikh Ahmad that he had best settle for what he could get, for his dreams of a vast subvention were fast fading.

His fears were confirmed on 22 December 1934, when his trusted friend, Colonel Dickson, the political agent, strongly advised him to sign the agreement which had been drawn up by the two companies. It had already been approved by the British government. It gave him a down payment of $170,000 instead of the half million dollars he had expected, and a yearly rental of $35,000 instead of $100,000. The payments were, moreover, to be in Indian rupees and were not to be backed by gold (as were those to Saudi Arabia, Iraq and Persia). The concession was to be for seventy-five years, and the price agreed as royalty per barrel of oil extracted was by far the lowest granted to any petroleum-producing country in the Middle East (twenty per cent less than the royalty rate for Persia and Saudi Arabia).

It was a salutary lesson to the Arabs of what happens when a country lacks independence and the companies competing for its petroleum decide to work together. But Sheikh Ahmad had no option but to sign, and he did so on 23 December 1934. Chisholm signed on behalf of Kuwait Oil Company, the new company which had been formed to exploit the new field, and Holmes on behalf of its associate and equal partner, Gulf Oil. Trusted friend Colonel Dickson witnessed the signatures. It was a happy outcome for the two negotiators, and in Kuwait's Anglican church on Christmas Day, as they filed into their pews, Chisholm and Holmes actually smiled at each other, and Chisholm's patrician face failed to show its customary wincing expression as the New Zealander's plangent voice was raised up in song. It was the season of goodwill, and none knew it better than these two.

The pre-World War II game of concession-grabbing was almost over in the Middle East, and in the Persian Gulf area there was only one likely area left where oil prospects were promising enough to bring the protagonists back to the table again. This was Qatar, a small sheikhdom just across the water from both Bahrain and Saudi Arabia, and therefore probably concealing beneath its rolling sand-dunes the same rich reservoirs of petroleum.

Stephen Longrigg, who was at the time a member of Anglo-Persian Oil Company's field staff, writes:

Meanwhile the Qatar prospect could not be followed up, and a negotiator was sent to Doha [the Qatar capital] to act nominally on behalf of Anglo-Persian so as to avoid confusing the old ruler with apparent rivalries, but in fact to acquire rights for the I[raq] P[etroleum] C[ompany].[7]

The British did not want, in other words, another of those

auctions which were so apt to delight the sheikhs by driving up the price of the concession, and when Anglo-Persian heard that both Standard of California and Major Holmes and his friends were planning to bid they speedily asked the Colonial Office to take action. For, of course, Qatar was yet another British dependency.

The Political Resident at Bushire felt it well to advise the Sheikh [of Qatar] to make the award to a British company and discouraged accordingly a tentative Californian demarche; he was aware also that Holmes and his Syndicate, who appeared with offers for the Qatar concession, were middlemen scarcely to be preferred to a fully competent operating company.[8]

But though everything seemed to be in Anglo-Persian's favour, the negotiations took time. The wily old ruler of Qatar refused to be rushed, and exasperated officials in London changed a polite and patient negotiator for one with less hesitation about invoking the name of the British government. It was not until May 1935 that the ruler, whom Longrigg describes as 'senile and suspicious', consented to award the concession to Anglo-Persian. They at once assigned it to the Iraq Petroleum Company, who, still surfeited with oil, were in no hurry to exploit it. Not until 1938 did they spud-in their first well.

Now the Middle East was divided up between the giants. Persia's oil was in British hands, as was Qatar and half of Kuwait. The Americans had moved over from Bahrain into the oases of mediaeval Saudi Arabia. A suspicious consortium of British, Dutch, French and Americans controlled the wells of Iraq.* There would be new rivals by the end of World War II, and new oilfields to conquer: in the deserts of Cyrenaica and Tripolitania, in the exotic emirates of Abu Dhabi, Dubai, Muscat, Oman and Dhofar, and beneath the seas of the Gulf. But for the moment all the productive as well as the most promising territories had been parcelled up. The rulers of the leased lands had had little to say about how it had been done, and their people nothing. There was nothing they could do about it, anyway: the leaseholders had powerful governments behind them to back up the terms of their tenancy. In any case,

---

* Baku had now been walled-off from the Middle East, and was still recuperating from the depredations of the Russian Revolution.

the Arabs had no technical knowledge or skills to exploit their riches themselves. It was a time to take the money (in the case of the rulers) and what labouring jobs were going (in the case of the people), to remember the Arab saying: 'The hand you cannot bite, kiss it.' And wait and watch.

PART TWO

The Operators

7 · Gusher

The first challenge to the supremacy which the Anglo-Iranian Oil Company had now established in the Persian Gulf came not from the Americans but from a company in whose policy and destinies Anglo-Persian had a considerable influence. Iraq Petroleum was a quarter-owned each by Royal Dutch-Shell, Anglo-Persian, France and a consortium of US oil interests. The company had been in operation in the new kingdom of Iraq long before King Faisal confirmed its concession in 1925, but as late as 1927 it had not yet come in with a discovery capable of even whispering a challenge at the British giant across the border of Persia. Most of the shareholders did not seem to be much concerned about that. Royal Dutch and Anglo-Persian were quite well satisfied with their operations elsewhere, as was Standard of New Jersey, one of the principal members of IPC's American faction. So long as no one else was able to exploit Iraq's oil, why waste money by hurrying. When the time was ripe and the markets were more favourable, that was the moment to begin spending on an accelerated programme.

Only the other American members of IPC's shareholding (Standard of New Jersey apart) were restive, anxious to press on, eager to find oil. This was their first, their pilot, venture in the Middle East and they needed to make a success of it if future undertakings were to be sanctioned by the directors back home.

By the spring of 1927 the news from Iraq, as far as they were concerned, was black, and not with oil. In April that year a well had been ceremonially spudded-in, with King Faisal invited to watch it, at Palkhana, in the Mosul district, but after drilling had gone down to 2,000 feet at this and one other well work was suspended. The limestone in which oil is trapped had not been reached.

Work then began at Tarjil, near Kirkuk, but all that came

66

up was salt water. In May the crews moved on to Jabal Hamrin, 'where a well at Khashm al-Ahmar was abandoned after encountering high pressure gas and heaving shales, and another at Injana was for similar reasons not taken below 3,500 feet. A well at Jambur, south of Kirkuk, met with minor shows of oil and gas. Results at Qaiyara, west of the [River] Tigris, were more significant.'[1]

It was all terribly disappointing. And expensive. As the meagre results came in, the members of the IPC board held an emergency meeting in London, and someone pointed out that 2,500 Iraqi workers, 50 British, and 20 foreign drillers were now operating in Iraq, together with domestic and office buildings, shops, stores, garages and vehicles. In a time of economic difficulties, it was all monstrously costly. A majority of the board seemed to be all in favour of cutting back in staff and expenditure and, at least for the moment, calling it a day. On the other hand, the minority American shareholders, the French, and Calouste Gulbenkian (whose fortunes depended at this time on the finding of oil in Iraq) were desperately eager to go on. It was finally agreed that there would be one last attempt to find oil, at the spot where the main drilling crews had established their headquarters. It was a dried-up river bed or *wadi* in the desert near Kirkuk called Baba Gurgur, and most members of the board were sceptical of the chances of anything worthwhile being found there. The company's chief geologist, Professor Hans de Bockh, had reported that there was 'an absence of the lagoonal phase in the tertiary sediment, which is the source bed of Asmari oil',[2] which meant that a viable field was unlikely. But the professor's opinion was strongly contested by the Americans in the syndicate, for they had sent out their own geologist, Doctor Richard Trowbridge, and he had reported enthusiastically on Baba Gurgur's prospects.

So reluctantly the board consented to one last try. On 30 June 1927, American drillers spudded-in the well and drilling commenced in the sickening heat of an Iraqi summer.

Now it so happened that there was something special about Baba Gurgur. About a mile and a half away (in between Baba Gurgur and the Kurdish city of Kirkuk, five miles farther on) a ring of flames about six yards wide and four feet high rose from fissures in the parched, stony black desert. By day the flames were invisible, and all that could be seen were trails of black

smoke rising into the sky. But at night, the flames could both
be seen and heard, and the locals said they 'licked like hyenas'
tongues and hissed like vipers'.

They were known (and still are known) as the Eternal Fires,
and they are said to have been there since Shadrach, Meshach
and Abednego marched through the 'fiery furnace' and the
idolatrous King Nebuchadnezzar cringed before them in mortal
fear of the wrath of God, burning continuously for five
thousand years. The drilling crews not unnaturally considered
the Eternal Fires a symbol and an augury of the success of their
latest well.

Drilling went on all summer, but as Beeby-Thompson, a
visiting oil engineer, afterwards wrote:

So little interest was taken in this test [well] that the drillers were
left much to their own devices, and although they repeatedly
reported encouraging showings of oil in the mud returns, little notice
was taken until cores showed that the basement bed [of rock] had
been reached, and it was considered useless to proceed further.
Orders were therefore given to cement the well for exclusion of
upper water and to make a test bailing; but on completion of that
operation the onus of testing was left to the American drillers.[3]

Shortly after midnight on 14 October 1927, a veteran Texas
driller named Harry W. Winger left the field compound and
joined his team of two other American drillers and a crew of
Iraqi helpers at the wellhead. The drill had now gone down
1,500 feet. Though he was not looking for miracles from the
final test which he and his fellow drillers had now decided to
make, Winger was too old a hand not to take precautions
against the unexpected. The rotary drill had already been exchanged for a percussion drill, to prevent unnecessary heat.
Now he ordered the Iraqis to bale out the casing, but told
them to leave 500 feet of mud at the base as a cushion against
any oil which might be, *might* be, surging around 1,500 feet
below. Beside him the steam engine which drove the percussion
drill began puffing away like an old-time shunting locomotive as
it worked the probing bit down into the rock below. It was a
cool night and a light breeze blew across the desert from the
direction of the Eternal Fires, bringing with it the pungent
smell of petroleum gas, a smell Winger had grown so
accustomed to that the air smelt dead whenever he moved out

of the oilfields. He remembered looking across at the fires and thinking what false prophets they were; if this well was a poor giver, as everyone seemed to believe, then he would have to be thinking about a new job.

After a time, he gave the order to withdraw the bit and clean the well of the cuttings which had been made, and it was then it happened. There was a noise, faint at first, deep in the bowels of the earth. It reminded Winger of a fast train approaching through a tunnel. It grew louder and louder, and at the same time mud and oil began to slop and glob and then gush on to the floor of the derrick.

Winger didn't wait to see more. He clanged the alarm bell and dashed across to the steam engine to douse the boiler, at the same time screaming to his Iraqi foreman, Sayed Nur Ali, to switch off all current at the electric switchboard.

Just before the lights went out, startled day-shift workers, alarmed by the bell, dashed from their bunks and were in time to see a shining black column of oil burst through the floor of the derrick like a genie out of a bottle and soar into the sky. Then all went black. They stood there, listening to the noise of the escaping oil, which was now an ear-piercing howl. They began to choke on the powerful stench of escaping gas. And then, from the blackness overhead, they felt rain falling on their heads and realized, when they touched it, that it was oil pouring down on them.

It took a little time for them to sense that the darkness was not total, and then they understood why. Over a humpy hill a mile away, the Eternal Fires still flickered against the blackness of the night. If the wind should turn and blow the erupting pillar of gas and oil that way, every man of them would be blown to bits – and perhaps the nearby city of Kirkuk with them.

It was then that Henry Hammick, the field manager, decided that what no man had done for five thousand years, now he must do. He ordered Alec Kinch, his English labour boss, to gather together men and all the spades they could lay their hands on. There was nothing they could do about the spouting well until morning light came, but they could make their way to the Eternal Fires. And in the next seven hours, with rocks and heaps of sand and rubble, they smothered out the flames of the Eternal Fires. And then, in the smoky, sulphureous light of dawn, they wiped the sand and oil out of their eyes and went back to the job of getting the runaway well under control.

By now the column of oil looked like a gigantic cypress tree bending in the wind against the pink morning sky. It was already two hundred feet high and growing. For ten miles around the sky rained oil, and in the camp itself it was already ankle deep and flowing down the *wadi* in an ever-growing stream.

Five men died in the fight to control the well at Baba Gurgur. Two American drillers wandered into a pocket of gas in a hollow near the camp and succumbed, as did three Iraqis who went in to try to drag them out. The reports in the records of the Iraqi Petroleum Company give a salutary picture of the appalling conditions under which the struggle was waged. The English manager of the company had travelled up by train and car from Baghdad as soon as he heard news of the emergency, and he was confronted by a nightmare scene. The oil was belching out of the well at the rate of 12,500 tons a day and the river of black crude flowing down the *wadi* was now 100 feet wide.

Sand and oil, coupled with the whistling wind, and the roar of the escaping oil [wrote Hilary Bull, the manager], made one wonder if it was not beyond man's resources to cope with the situation. No man could work for long under such conditions and ten minutes seemed to be the average endurance ... With gas masks on, and oil dripping from their hats, the drillers worked feverishly in the cellar [of the derrick]. The man actually at the bottom of the cellar fixing the tie rods under the flange had a rope tied round him so that he could be hauled up if overcome by gas. All day long men were collapsing and being hauled up unconscious to be revived outside in the comparatively fresh air. Some of the men were gassed two or three times a day, and yet staggered back to their jobs.

No one had believed enough in the potentialities of Baba Gurgur to make provision in the event of a gusher, so makeshift *bunds* had to be built in the sand to contain the spurting oil as the drillers wrestled to contain it at the wellhead. The first *bund* filled up in one afternoon, and there was a frantic race to get a second and bigger one dug before it spilled over. Six *bunds*, each larger than the one before, were erected. These were days before oil companies thought too deeply about pollution, but everyone knew that the River Tigris was not far away and that it would be a disaster of the first magnitude if the flow of oil reached it. Finally a site was chosen seventeen miles from the wellhead and 800 men from a local tribe were hired to make a reservoir there. It was to be 6 feet high and 700 feet long, and it would hold 200,000 tons of oil. The menacing black river, slopping over from one *bund* to the other, was only three miles away from this last ditch when the drillers finally got what is known as a Christmas tree (a new wellhead valve) into place and brought the filthy roaring menace under control at last.

It was three p.m. on 23 October 1927, nine days after the oil and gas had first burst free, and the sudden silence was uncanny. Exhausted men flung themselves down on the oily ground. Someone dared to light a fire and make some tea. Bottles of whisky were broken out of the compound's emergency store. And as they sipped through oily lips, Hilary Bull saw a car approaching the compound from the direction of Kirkuk. It was a messenger from the local provincial governor demanding the manager's presence at once.

'And now,' the governor said, some hours later, his tones icy with fury, 'now that you have your oil, will you please relight the Eternal Fires as urgently as possible? For nine days my people have been without them, and it has made them profoundly uneasy.'

The flames were flickering over Baba Gurgur a week later. The Eternal Fires were in business again, and the Iraq Petroleum Company was launched on a veritable sea of oil. And many were the blessings that flowed therefrom.

8 · Domes for Sale

In the summer of 1934 Max Steineke waded ashore at Jubail, on the Gulf coast of Saudi Arabia, from a small Arab dhow which had brought him from Bahrain on the last leg of his journey from the United States. He had been appointed geologist-in-charge of the new American concession in Al Hasa, and on his way over he had read closely the reports of the pioneer party of ten Americans who had been in the area for over a year. But their sparse prose had hardly prepared him for the realities of the situation. The heat was blinding, and there was no relief from it even wading waist-deep in the waters of the Gulf, for the temperature of the sea was hot almost to the point of simmering. Nor was his mind made easier on this final few hundred yards of his long journey by the sight of a long sea snake, its mouth open and its wicked-looking teeth bared in what appeared to be a snarl, undulating towards him. He shied violently like a frightened horse and dropped his type-writer into the water, whence it was retrieved by an Arab crewman, who almost immediately thrust him violently to one side, so that he fell forward, ducking his head in the water. A small school of pink jellyfish floated by, their delicate tentacles spinning like ballet dancers' arms a few inches from his nose. It was only when he reached the shore, angry and soaked, that it was explained to him why the Arab bearer had been so rough. In this part of the world you ignored sea snakes, which, no matter how fierce they looked, seemed to leave human beings severely alone, but you took immediate evasive action when confronted by pretty pink Portuguese men-of-war, for the lightest caress from their delicate fronds could give you a rash that would agonize for weeks.

The world was one third of the way into the twentieth century, but you would never have guessed it from this part of

72

Arabia. In some ways it was almost like the Old West of a century before, except that the wandering tribes were bedouin instead of Indians and the herds they tended were camels instead of buffalo. In villages like Jubail and Oqair and in the nearby city of Al Hufuf, centre of the Al Hasa oasis, the local emirs received their writs from King Ibn Saud and ruled like feudal princes. The Americans were tolerated because they would find the oil that would make all of Arabia rich, but even so they were viewed with intense suspicion by the emirs as infidel intruders come to corrupt and suborn the tribesmen. No latitude was allowed them. On the beach at Jubail, beside Steineke's luggage, lay crates of tinned food which had been sent from America via Bahrain as rations for the American crews; but the emir had decided that they were liable to heavy import duties, and though the Americans pointed out that this was contrary to the terms of the concession agreement, the crates simmered in the sun, and the crews subsisted on a diet of rice and teeth-breaking camel, goat or occasional mutton. In consequence, most of them were suffering from stomach pains, boils and spots-before-the-eyes, which may have been psychosomatic or may also have been due to the water, which tasted of camel urine. All of them walked and moved their limbs with difficulty, for they were martyrs to prickly heat in the most unfortunate places, and this they put down to heat, sand, and the rigours of the country.

In the circumstances, Steineke found them a surprisingly cheerful and optimistic team, and that suited him, for his temperament was very similar. Since he had left Stanford University he had worked in all sorts of difficult places – South America, Australasia, Alaska – and he had learned to put up with physical hardships; and though he disliked the heat, the flies and the stench of his new location, the people and the landscape were something else again. He was a bluff, laughing man with a rich vocabulary of swear words, and ordinary folk (if not emirs) soon succumbed to his friendly approaches. As for the desert and its hills and formations, it looked promising for both work and pleasure. The formations of the *jabals* or hillocks suggested likely petroliferous indications; and both the sea and the sand-sea were still rich with enough fish, birds and other fauna to satisfy his passion for hunting, for he was both a crack shot and an expert big-game fisherman.

The Aramco* team lived in 1934 in a *barrasti* (a sort of bunkhouse made from woven palm leaves) between the tiny pearl-fishing village of Jubail and the main town of the area, Dammam. They had a couple of trucks in which they bounced over the desert with their supplies (there were no roads) and otherwise got around on donkeyback, like the Arabs. They also had a plane, and that had caused more trouble than anything. Another clause in the concession agreement said that the company could use up to two planes for area mapping and reconnaissance, and a Fairchild monoplane had been shipped to Egypt, and flown in from there by its pilot, Dick Kerr, by a complicated route via Palestine, Iraq and Iran. Unfortunately, someone had forgotten to inform King Ibn Saud that it was coming; it was a time when the king was in the midst of one of his recurring quarrels with the imam of Yemen, and suspicious that his adversary was receiving secret aid from Iran. Kerr chose this moment to fly in his Fairchild; and when a Saudi radio operator heard the plane talking to an air control station at Bushire, Iran, before coming in to land at Jubail he signalled the alarm to the king in Jiddah.

It was the job of the oil company representative in Jiddah, Bill Lenahan, to straighten out these difficulties, and he had managed to convince the king that Aramco was not secretly aiding an Iranian invasion. The plane had been released from impoundment but its flights were restricted to coastal reconnaissance, and the emir of Al Hufuf had given orders to his irregular troops to open fire if any plane should fly over the oasis.

The pioneer team had already staked out a number of likely sites for test oil drillings up and down the coast, and particularly between the coastal town of Dammam and a series of rocky hillocks or 'structures' known as Jabal Bari and Jabal Dhahran. Steineke told one of his assistants, Tom Koch, that it looked as if he had an easy task ahead of him. The theory was that the 'structures' out of which Bahrain's oil had come continued under the sea and on to the Saudi mainland.

The problem seemed merely one of cruising about the desert and

* The company was known as Casoc (California Standard Arabian Oil Company) in these days and did not change its name to the Arabian American Oil Company until 1944, but I have used the present-day title for clarity.

finding jabals—an easy matter [wrote Steineke later]. In the lights of
what we now know about the geology of Arabia, these early
endeavours to bring to light the secrets locked away under the sand
seem very amusing. But they were done in deadly seriousness in
those days. Krug Henry and Bert Miller (two oil company employees)
had found many such jabals during their trip to the west in the spring
of 1934. Muhamet Taweel, the emir of the Hasa Province, became so
eager to see the oil produced that he caused many anxious moments
with his trying to promote the drilling of wells on each jabal, so sure
was he that each one would prove to be a field. At the first opportunity, therefore, we drove out to investigate Jabal Bari in detail.[1]

Petroleum is usually found in so-called 'sedimentary' rocks
which were once the bed of ancient seas, and in the 1930s what
the geologist looked for were evidences of fossilized shells in
the rock formations. These discovered, he went on to search
beneath them for the 'cap rocks' in which oil, formed from fatty
organic matter masticated on the sea bed by bacteria, is trapped
by the movement of water and the pressure of the earth. Jabal
Bari seemed promising in that it was a troughlike fold in the
rock formation, but when Steineke chipped out a few hunks
of rock he discovered small, round, buttonlike fossils that
belonged to the Miocene period, only nineteen million years
old, and too young for his purposes. He wanted Eocene beds like
those at Bahrain, at least fifty-five million years old.

After a series of probes which had proved both disappointing
and confusing, he moved westward about twenty-five miles to
a spot which the pioneers had named Button Bed Hill. Steineke
had lumps of rock in his truck which had fossilized oysters, sea
urchins and other small echinoids locked in them – signposts of
Eocene – but each time he dug down the formations were thin.
He even came across one outcrop of Cretaceous rock, filled with
clam skeletons and giving evidence of an age of more than
seventy million years. But each subsequent probe, at Button
Bed Hill, at Jabal Dhahran, and over a number of promising
domes or 'caps' which had been mapped out from the air all
proved, after weeks of weary test drilling, to be sterile or un-
promising. Some wag put a cartoon up on the mess wall of
Steineke weeping tears upon the jabals and a notice over it
saying: 'Domes for Sale', but nobody laughed very much. The
wildcatters who were working under Steineke's orders from a

desert camp outside Dammam had the worst job in the world, working, soaked with sweat, in a temperature that rose to one hundred and twenty degrees by noon, often in shrieking sandstorms, and aided by bedouin workmen whose language they did not speak and whose minds were unattuned to this kind of labour and these newfangled tools.

Now the bets started pro and con as to whether or not even Dhahran would be worth fifty cents as an oil field [wrote Steineke]. The year before Buck Miller, Krug Henry and Soak Hoover had mapped Dhahran as a bona fide structure similar to Bahrain which was of course a producing field. But now the theories we had constructed based on these facts seemed to be going more and more awry. There was something very unorthodox about the whole geological set-up . . . The whole situation was causing us sleepless nights when we argued and discussed all these weighty problems trying to solve them in the light of our misconceptions.[2]

Not that there was any despair around (except in Jiddah, perhaps, where King Ibn Saud was waiting for the oil, and the money), for the company's home office in San Francisco showed no signs of backing out of the project, and was, in fact, continuing to beef up its staff in Saudi Arabia both in men and materials. Aramco may have committed some serious errors in the course of its career, but, unlike other companies, it has never operated as an absentee concessionaire. It has always put its senior men into the hot spots, and drawn its top directors from the field; and working beside Steineke were Fred A. Davies, who was to become Aramco's president, and Floyd Ohliger and Tom Barger, who retired as executive vice-presidents, all three of them geologists and engineers. San Francisco gave every evidence of backing the Saudi project or going bust, and so far as Steineke and his crews were concerned, go bust they would not.

But things got edgy at times. There was a crisis when one of the drillers clumped a Saudi workman for stupidity and insolence, and was ordered out of the country by a furious King Ibn Saud. The drillers said they would all go if he went, so he stayed, but relations with the local emir and his officials were not improved. Henceforward, Americans were roughed-up by the police for the slightest infraction of local rules and customs, and were finally confined to their compound or camps. At the end of 1935, Max Steineke having decided to stake his all on the so-called Dammam Dome, King Ibn Saud's eldest son, Saud, came across from Al Hufuf to turn the wheel on Dammam Number One, which had gone down to 3,200 feet and was producing oil. But the royal hand worked no miracles and the well whimpered away after a promising start. As a compensation the company lent the king £15,000, for his wives and other extravagances had put him heavily into debt again.

'Still going for bust,' as Steineke termed it, the company ordered a whole series of test wells to be drilled on the Dammam Dome, and the geologist realized that the crucial phase of his professional career was now looming before him. If the tests failed, not only would Aramco be bust, but he would be too.

It was now 1936. Dammam Number Two, after a good start, suddenly vomited salt water. Number Three brought up a sluggish sulphureous oil so viscid that all they could do was use it as covering for the desert roads they were building. Number Four went down to 2,318 feet and stopped there, on 18 November, dry. So did Number Five, at 2,067 feet. Number Six was never even spudded-in, but simply had a derrick erected over a probing hole, and was left there.

It was left there because Max Steineke had decided to put all his money, all his prospects and those of his colleagues and his company, on one final well, Dammam Number Seven. He was so confident of success (or pretended to be) that after seeing it spudded-in, on 7 December 1936, he departed with an engineer, Floyd Meeker, on a trip across the Arabian Desert to Jiddah and back. The company was encouraged by his optimistic predictions, to such an extent that they gave the go-ahead to a programme for sending wives to join their husbands in the Gulf, and ordered a married-quarters construction scheme to be started.

Fortunately for all concerned, seven is a lucky number in the oil business.

On 31 December 1936, Dammam Number Seven 'blew out' and sent rigging and drill high into the air. The drill was down by then to 4,535 feet, and New Year's Day was spent pouring great loads of mud into the well to 'kill' a jet of gas which was blowing out at the rate of 30 million cubic feet every twenty-four hours. It was gas only, no oil with it, and at that time no one was interested in gas, only frightened of it.

For the first time, home office was frightened too. Had

Steineke guessed wrong? Should they try somewhere else? The geologist, back from his trans-desert trip, was confronted with these problems and asked for an immediate answer. He agonized through a sleepless night and then told them to press on with Dammam Number Seven.

On 4 March 1938, oil began to flow out of the well at the rate of 1,585 barrels a day. Three days later the flow had increased to 3,690 barrels. By the end of April the well had produced 100,000 barrels – and Steineke predicted that it would go on producing.

They were into an oil cap, and a commercial field.[3]

But no one told King Ibn Saud yet.

By the time Dammam Number Seven put the Americans in the oil business in Saudi Arabia a change had come over the Aramco encampments, and it could be truly said that the pioneering phase was over. For by mid-1937 the first five American wives, and three of their children, had arrived and were installed in wooden, California-type bungalows which had been erected amid the scrub and wastes of sand under the slopes of Jabal Dhahran. No one imagined then that these were the nucleus of what, thirty-five years later, would be the green garden-city of Dhahran, capital of the most prosperous oil consortium in the world.

There had been oil company wives living in Iran, Iraq and Bahrain for years – the first British wife was at Masjid-i-Sulaiman, in Iran, as early as 1910 – but all these territories at that time were under British control or influence, and years of colonial rule had accustomed the locals to the presence of European women. Saudi Arabia was different. It was a closed society where a camel was more valuable than a woman, where the king did not even bother to count up how many daughters he had sired, and where Christian females were despised even more than their Muslim sisters. Ameen Rihani, a devout Christian, was shocked by a story told to him by one of Ibn Saud's ministers about a beautiful young Armenian Christian girl who had been bought as a slave by the King in the 1920s. He then gave her away to his minister 'after he had entered her – one night only'. The minister went to her that night in the room to which she had been assigned.

She held down her head after she had seen me [the minister said]. And I—never in my life have I seen or heard of such beauty, such august beauty. Her skin? White as alabaster. Her hair? Like cataracts of melted gold. Her lips? Red as a pomegranate seed. Her forehead? Lofty and glowing like the dawn. And those honey-coloured eyes, so soft, so demure, so appealing to the honour of man . . . I sat before that image of beauty like a child—I tell you, like a child—and I felt the flame of shame upon me. I was ashamed to touch her, or even to speak to her. I got up and walked out of the room.

And then what did he do? Rihani asked the minister.

'And on the following day,' he replied, 'I sold her to a man from Kuwait for 400 rials. Aye, wallah, only 400 rials.'

Faced with a rigidly misogynistic society, it is not surprising that someone suggested that the American women should consider wearing veils. Max Steineke's wife, Florence, settled that by saying:

'I'm damned if I'm going to cover my face for any man in the world. It may not be the face that launched a thousand ships, but those Arabs are going to have to look at it.'

The women did decide, however, not to wear slacks or the shorts they had brought with them from the States, and they reluctantly packed away their bathing suits. Then they trooped out into the desert and posed under the derrick of Dammam Number Seven for the official photograph of the first five Dhahran wives and offspring: Marilyn Witherspoon, Florence Steineke, Nellie Carpenter, Patsy Jones and Edna Brown, the Steinekes' two daughters, Marian and Maxine, and Mitzi Henry, all with chins up, grinning defiantly into the sunlight, unveiled.

For the first few months life must have reminded them of those stories from American folklore of life in the cantonments on the plains, for they were under orders not to leave camp, and they sweltered in a square of desert not much more than two miles by one, and, appalled by the blazing heat and the driving dust, wondered whether they had saved their pride at the expense of their complexions by refusing to wear veils. But when Dammam Number Seven came in, everyone, even the Saudi authorities, seemed to relax and things got better. As proof that they recognized the permanence of the operation, home office shipped in air conditioning equipment for the family bungalows. Shortly afterwards, Max Steineke took his wife and Nellie Carpenter on one of his forays into the desert. They dressed in Arab clothes and visited two bedouin encampments, where they spent much time struggling through sign language and gales of giggles to make themselves understood by the bedouin women. They came back vastly enlivened by what they had seen, and soon all the wives had their bedouin robes and were travelling the desert in them. And from now on they took ointments and surgical spirits and bandages with them, to tend to the sores and the wounds which they found among the bedouin women and children.

The more they saw of the life around them, the more excited, and, at the same time, the more appalled they became. It was the beginning of a love-hate relationship with Arabia which has endured among the Americans to this day. They were entranced by the vast, moody beauty of the desert, its rolling seas of golden dunes, its flat, black, calm infinities of shale and cinders, its sudden, rearing mountain peaks and ranges shaped like battleships and toadstools and fairy castles. They organized their husbands into expeditions to explore the caves and the blood-red fissures of Al Mubaraz and the hot springs and old Turkish forts of Al Hasa. Their children played with the Arab children and exchanged words of each other's language. They went out with the pearl fishers of Darin Island and gawped at the divers who went down to the sea bed with goatskins to bring up fresh water from the springs which bubbled into the bottom of the Gulf.

But the backwardness and suspicion and superstition depressed them, and the cruelty made them indignant. How could they keep silent when they saw an emaciated child, covered from head to foot in sores, each scab clotted with flies, crawling among the dungheaps in an Arab village? Florence Steineke penetrated into the black women's tent of a bedouin tribe and found a young girl with great open angry sores across her chest, into which an older woman was rubbing a noxious fluid. It turned out that the girl had been cauterized with a branding iron against the tuberculosis from which she was suffering, and now the wounds were being treated with the urine of a female camel. Hot irons, on the face or the heel, were favourite remedies for stomach ache; needles were used on the lenses of eyes with cataracts or trachoma; and the vagina of a newly-delivered mother was rubbed with rock salt immediately afterwards, to contract it in case her husband, being dissatisfied, turned to another woman. And that meant that next time she was pregnant, the scarred vagina would fail to stretch and the woman would agonize until she died.

These American women, warned by their husbands not to 'upset the apple cart' by making too much fuss, organized medicines, and, once nurses and doctors began to appear, set up clinics to give practical medicine and surgery to the pathetic victims of Arabian disease and ignorance. Sometimes it seemed that every child and every woman they saw had a badly-set limb, a diseased eye, yaws on the face, or was suffering from malaria, tuberculosis or diarrhoea. They did what they could to alleviate the suffering and, in fumbling but increasingly fluent Arabic, offered to the unfortunate females of Arabia what practical advice was possible under the social and physical conditions of this mediaeval land. But when they came to see a mother after she had given birth, and, asking her about the baby, heard her say 'It was nothing', they wept. For that meant that the woman (who was often a girl of twelve or thirteen) had borne a girl, and it seemed so monstrously unjust and unfair.

And what made them flinch, too, was the implacable brutality of Saudi Arabian justice. And that they could do nothing about. The first time that an American wife, driven to distraction by the petty pilferer of her household goods and cashbox, reported the thefts to a local official, she lived to regret it. The culprit was discovered, a Saudi houseboy, and speedily found guilty by the imam at Dammam. It was more than a month before she saw him again, and during that time no male member of the camp would talk about it, would tell her what had happened to him. Then one day she saw him on the outskirts of the camp, and on the end of his arm where the hand had been was a festering stump. Thieves had their hands ceremonially chopped off in public, and only because Floyd Ohliger, the general manager, had interceded, had she not been summoned to watch it. They all began to realize that they lived in a biblical, an-eye-for-an-eye society, where the relatives of a man, woman or child killed by an automobile could demand the life of the driver in return, could appoint the executioner, and decide the method of dispatch. They could be bought off, of course, and persuaded only to demand minor punishment, but that meant that it was only the poor who suffered.

It so happened that the emir of Al Hufuf was a stern up-
holder of the Koranic law, and was rigid and ruthless in his
punishment of sinners. Word of his brutalities spread through
the province, and he was feared far and wide. Phil McConnell,
a production man, came back from a visit to Al Hufuf to fill
them with horrid fascination as he recounted his unwitting
attendance at the public execution of a poor wretch who had
been caught committing adultery not only during the day, but
in Ramadan, the time of religious abstinence:

They had shaved off his hair and his beard, and they brought him
on to this platform with his hands tied behind his back. The poor
guy looked surprisingly calm and resigned. There were two
executioners, big Negro slaves with chests like barrels. They made
the man kneel down and one of the Negroes, he had a huge curved
sword in his hand, and so did the other, went and stood in front of
him. And suddenly he started dancing! He waved his sword above
his head. He spun around in his sandals, and the crowd cheered him.
As for the poor guy on his knees, he was watching, fascinated. Sayed
[his interpreter] whispered to me that that was the idea. While his
attention was distracted, the real executioner was getting ready
behind his back. Suddenly he moved his own sword, quick as a flash,
and I saw the end of it prick the back of the man's neck. His head
jerked back—Sayed said that was to stiffen his muscles—and then
so fast all you could see was a blur of silver, the sword came down
and the head was off and rolling on to the platform, and blood was
gushing out of the neck.[4]

It was no comfort to have it pointed out, as Harry St John
Philby seldom failed to, that this stern form of justice kept
Saudi Arabia free from thieves and murderers at a time when
elsewhere in the Middle East no one was safe from robbery and
violence. Faced by the horrid thought that the thief would lose
his hand, or even his arm, if reported to the authorities, the
Americans began overlooking petty thefts, and thus encouraged
their increase. At one moment they even persuaded Floyd
Ohliger, who was now on close terms with the king, to ask him
to allow the culprit's severed stump to be treated immediately
after the ceremonial severing, against the flies and infections to
which it was exposed. But Ibn Saud would neither allow this
nor the smoking of kif against the pain. His attitude was: if a
man is being punished for a crime, he should feel his punishment. And that was the end of it. The wives were urged not to

insist; and many was the petty theft that went unreported at Dhahran in the years to come.

It was not until October 1938 that King Ibn Saud was officially informed that a viable oilfield had been found at Dhahran,* and by that time Dammam Number Seven was not the only well among the jabals. Others were being spudded-in all around the original bonanza. For Ibn Saud it was a welcome relief, for, of course, he and his administration were heavily in debt once more. He was so thankful at the news that he informed Floyd Ohliger and Bill Lenahan, the two Aramco representatives, that he had refused to grant new concessions to certain applicants who came hurrying to Jiddah, jingling gold, the moment the news of the Aramco strike leaked out. One of them was Stephen Longrigg of IPC who retired voluntarily once he heard that Aramco had an option over the concession he was seeking. The German ambassador to the Middle East, Fritz Grobba, came in from Baghdad to put out feelers on behalf of German and Italian interests.† So did the Japanese. All of them, the king informed the Americans, he had turned down, even the Japanese, who had made him a 'colossal' offer.

But later the Americans had another meeting, this time with Ibn Saud's smooth and expert finance minister, Abdullah Sulaiman, and he made it politely clear that what had been refused the Japanese and Germans he expected Aramco to take over, at a price above what the others had offered. It was obviously going to be a vastly greater sum of money than the company had paid for the original concession.

For the moment, however, the bargaining over new concessions was postponed for a ceremony. The king announced that he would journey to the Gulf coast to pay his first visit to the company's oilfields and installations. He arrived on 30 April 1939, after a desert trek of a thousand miles with five hundred cars and two thousand slaves, servants, wives and concubines. On the morrow he would declare open a new Aramco port. Miraculously, a city of bedouin tents arose beside the California-style bungalows of Dhahran, and at sunset the Americans watched in awe as the voice of the muezzin rose from a nearby mosque and the giant monarch marched ahead

* The US oil township was officially named Dhahran in February 1939.

† There were rumours that Standard of New Jersey, which had considerable interests in Germany, was tied up in the negotiations.

of his followers into the desert and then the whole concourse fell on their knees in the sand and bowed their heads westwards to Mecca for the last prayers of the day.

This part of the Gulf coast was a vastly different place now from the bare, barren coastline Max Steineke had first seen in 1934. A pier had been built for ships to unload at the fishing hamlet of Al Khobar. Out on the isthmus, through the pale-blue shallows a sea passage had been dredged for ocean-going tankers. A tank-farm had been erected and underwater pipelines led out to floating anchorages. The first tanker to tie up at the pumphead, the D. G. *Schofield* was in position and waiting. The loading port was called Ras Tanura, and the Americans were proud of it. Telegrams of congratulation were read out from William Berg, president of Standard of California, and Torkild Rieber, chairman of the Texas Company.* Ibn Saud and his finance minister, Abdullah Sulaiman, were presented with new automobiles (to the king a Cadillac, to Abdullah a Chrysler). Floyd Ohliger and Bill Lenahan got gold watches from the king in return, and two golden daggers elaborately carved by the craftsmen of Al Hufuf. Then Ibn Saud, from a gilded chair, turned the wheel that set the Ras Tanura pumps in motion, and the D. G. *Schofield* began to fill with crude.

Ten days later, back from an official visit to the ruler of Bahrain, whose oil revenues he had always envied, but whose good fortune he could now match, the mighty king threw a great banquet for four thousand people on the sands of Dhahran. More than a thousand sheep were slaughtered, and few of the three hundred Americans who now made up the company's complement escaped having to swallow a sheep's eye thrust upon them by their hosts. The party continued into the small hours, lit up by the gas flares from Dammam Seven, Eight, Nine, Ten, Eleven and Twelve, now busily pumping oil into the tank-farm of Ras Tanura and gold into the pockets of Aramco and King Ibn Saud.

After all the setbacks and disappointments, it seemed too good to be true. And it was.

On 3 September 1939, Britain and France declared war on Germany, and World War II began. The effect on the fortunes of Aramco, on Ibn Saud and on Saudi Arabia was disastrous.

* The two companies which, at that time, controlled Arabian oil.

9 · Skulduggery in Kuwait

The sheikhdom of Kuwait, that wedge of desert at the head of the Persian Gulf, between Saudi Arabia and Iraq, is only seventy miles broad by eighty miles long, but beneath its sands is a veritable sea of oil. There are so many indications of its presence that it would seem difficult to sink a well and not find it. But in 1936 Anglo-Iranian put down its first well, on its own behalf and that of its US partner, Gulf, and missed the oil by miles.

Was it deliberate? The independent government of Kuwait is at the moment (1972) hoping to confirm or confound its suspicions on this point by commissioning an official history of the Kuwait concession and its subsequent operation by the Kuwait Oil Company. It is being prepared for them by Mr Archibald H. T. Chisholm from his retirement in Eire, and the fact that Chisholm was Anglo-Iranian's chief negotiator in Kuwait should give him access to material which is unavailable to other historians.

The agreement on the concession which Chisholm and Frank Holmes (for Gulf) extracted from Sheikh Ahmad was a hard bargain for the Arab ruler and his people, and one they lived to resent. But the agreement which Anglo-Iranian made with Gulf, once they decided to share Kuwait's oil, was something in the nature of a triumph for the British company, also. It will be remembered that Anglo-Iranian did not really want to see any more petroleum produced in the Gulf, or anywhere else in the Middle East, for each new oil source prejudiced the agreement to control the markets which had been made at Achnacarry by Standard of New Jersey, Royal Dutch-Shell and Anglo-Iranian. These companies would have preferred Iraq and Iran, where they owned the oilwells, to monopolize the Middle East's share of the world fuel markets. The discovery

85

of oil by Standard of California (a company outside the charmed circle) at Bahrain had been a jolt to their plans, and they feared the effect of Saudi Arabian operations. So when Anglo-Iranian forced Gulf into a partnership in Kuwait, the terms the British wrote into the agreement with the Americans were harsh and restrictive. Both parties pledged that Kuwait oil would not be used to interfere with or damage the world market, and that they would keep in constant consultation over where the oil was sold. And then came this clause:

> The parties have it in mind that it might from time to time suit both parties for Anglo-Persian to supply Gulf's requirements from Persia and/or Iraq in lieu of Gulf requiring the company to produce oil or additional oil in Kuwait. Provided Anglo-Persian is in a position conveniently to furnish such alternative supply, of which Anglo-Persian shall be the sole judge, it will supply Gulf from such other sources with any quantity of crude thus required by Gulf provided the quantity thus demanded does not exceed the quantity which in the absence of such alternative supply Gulf might have required the company to produce in Kuwait . . .[1]

This gave Anglo-Iranian a position of power over its US partner both going and coming, since it was the British company which would provide the manpower and decide where the Kuwait wells should be drilled, and it was the British company which would have the choice ('of which Anglo-Persian shall be the sole judge') of deciding how much oil should come out of them once they were in operation. Anxious as the company was to keep up the flow of oil from its main source of supply, in Iran, it is not surprising that Anglo-Iranian was in no hurry to bring Kuwait into the game. The mystery remains how and why Gulf allowed them to get away with it, anxious as the Americans were for the stake in Kuwait oil.

There was one obvious place to drill for oil in the territory, and that was at Burgan. That was where fires had burned and bitumen had seeped for generations. It was the *wadi* in the desert of which the ruling sheikh, Mubarak al Sabah, had written to the British as long ago as 1913, promising to show a visiting Royal Navy admiral 'the place of bitumen in Burgan and elsewhere,' and pledging himself, if oil was obtainable therefrom, to keep the concession for it British. Frank Holmes, the Gulf negotiator, had always envisaged Burgan as the spot where he would commence drilling operations.

Anglo-Iranian geologists decided otherwise, and picked a remote spot named Bahra where there were, it is true, evidences of bitumen seepage, but nothing to compare with Burgan. Any doubts expressed about the site were demolished by self-confident company experts, and the ruler, Sheikh Ahmad, the new British political agent, Colonel A. C. Galloway, and a host of dignitaries were invited to the ceremonial spudding-in of the first Bahra well in May 1936. It was an augury for the future that the ceremony took place in one of the sandstorms for which Kuwait is notorious; and as guests peered, hot and sticky with sand, through the brown, swirling mist at the shadowy figures spudding-in the new well, only the thought of the oil to come made it bearable. But weeks went by and no oil strike was reported. Months passed, and though the drill bit deeper and deeper into the rock, there was nothing. It was 1937 before the drill was pulled out at 8,000 feet and the glum announcement was made that Bahra was dry.

Now everyone presumed that the drilling crews would move to Burgan at last. Sheikh Ahmad, short of money to pay for his last cabaret girl from Beirut, behind with his doctors' bills (he was rumoured to be suffering from syphilis), angrily demanded of the political agent, Colonel Galloway, the reason for the delays. There were whispers in the bazaars that the oil company was deliberately stalling and agitators whipped up enough anti-British feeling for a demonstration, which was quickly suppressed. But Doctor Fritz Grobba, the Nazi minister in Baghdad, busily subsidizing anti-British movements, was interested enough to send a courier down with money for the dissidents.

And still the British company held off from Burgan. Instead it announced that the whole of Kuwait territory was to be thoroughly mapped and geologically surveyed by specialists, and an elaborate (and some said time-wasting) operation began which lasted for nearly a year. Only at the end of it did the British announce that, after all, they had come to the conclusion that there was one spot which was more likely than any other in Kuwait to produce viable quantities of oil. Burgan.

It was there that a well was spudded-in on 16 October 1937, with no ceremonials this time. Promising oil shows were

87

reported when the drills reached 3,400 feet, and in April 1938, the bit cracked through a cap of rock and plugged into a vast swirling cavern of high-pressure gas, sand and oil, all mixed together. The clocks on the gauges spun like spinning tops, and it looked for a time as if the unleashed elements were going to blow drill and earth-crust sky high; only by desperate efforts was the pressure (greater per square inch than the strongest city firehose) brought under control.

Anglo-Iranian's joy was muted, but two other wells were at once put in hand, and by mid-1938 the two partners, and Sheikh Ahmad, knew that they were in possession of a phenomenal oilfield. It lay twenty-eight miles south of Kuwait town, fourteen miles into the desert from the Gulf. The field was pear-shaped with the stalk-end pointing northwards, and it was at least fifteen miles long. No one was yet aware that Kuwait was, in fact, sitting on one of the largest reservoirs of oil the world has ever known, but everyone realized that Kuwait's petroleum resources would henceforth alter the balance of oil production in the Middle East.

Once it went into production, that is. Unfortunately, Anglo-Iranian was not yet ready for that. First the Munich crisis in September 1938, and then the fraught, eve-of-war atmosphere of 1939 were given as excuses for holding back. There were more urgent priorities, it was pointed out, and the material was just not there to outfit a new oilfield. While Aramco, farther south in Saudi Arabia, built Ras Tanura and began ferrying tankers to the Bahrain refinery with regular loads of oil, Kuwait, with three wells ready to go, stagnated.

Sheikh Ahmad, short of money, aware of restiveness in the territory – for it was the worst of times in Kuwait, its pearl-fishing industry coming to an end, its petroleum wealth un-exploited – pleaded with Anglo-Iranian to get going. There were more riots from the volatile Kuwaitis to back him up, and the fact that they had been subsidized by the emissaries of the Nazi envoy, Dr Grobba, did not affect the validity of Kuwaiti discontent. It was a difficult moment for Anglo-Iranian.

Then Britain declared war on Germany, and let the company off the hook. Nothing more could be done, they told the sheikh, until Germany and her allies had been defeated. Until then, the oil of Kuwait must stay in the ground.

'We have been cheated!' cried one Kuwait Arabic newspaper. 'We should never have allowed Kuwait's lifeblood to fall into British hands. It would have been better if we had handed it to Germany to exploit.'

The newspaper was suppressed, but the words stayed like irritating grit in the minds of most Kuwaitis, and they would create difficulties for the oil companies and for Britain in the years to come.

PART THREE

The Arabs Hit Back

10 · Someone Else's War

In October 1940 a small fleet of Italian bombers carried out the not inconsiderable feat of bombing the oil installations of Bahrain, after a flight from the Mediterranean and on across the Red Sea and the Rub al Khali (the Empty Quarter of Arabia) to a base in Eritrea. The results the raid achieved were more psychological than physical, for though the bombs missed their target Bahrain's total inability to react against the raiders rubbed home for every Arab the cold facts of the situation. Like her one-time ally, France, now crushed by her German and Italian enemies, Britain was crumbling helplessly into defeat before the inexorable might of Nazi Germany. It would be untrue to say that this prospect alarmed most Arabs and Iranians, disturbed though they might be by the results of it. Britain had her Arab friends in the Middle East, of course, but even among those who admired her or respected her, there were few who loved her. In each of the countries where Arab kings, emirs or sheikhs were the titular heads, but where the real power lay with the British envoys, high commissioners or political agents behind them, there were active cells of dissident officers or politicians scheming and dreaming of the day when they could be driven out. There was the young officers' group in Egypt led by Gamel Abdul Nasser and Anwar al Sadat. In Iraq there was the fiercely anti-British army clique of colonels known as the Golden Square. In Kuwait there was a militant political revolutionary group called Al Shabiba (the Followers). The writ of the British colonial office did not run as far as Iran, but the oilmen of Anglo-Iranian had long ago turned the south-west of the country into a British enclave, and the ruthless and authoritative shah-in-shah in Teheran, Reza Khan, made no secret of his resentment of them, and of his contempt for the failure of British arms in the war. The hatreds and antipathies of the masses were fanned each night by the exiled Mufti of Jerusalem, broadcasting anti-British diatribes from Berlin, and by the Arab emissaries of Doctor Grobba, who had now moved to neutral Turkey but kept his lines open to Baghdad.*

If trouble came, the oilmen had no doubt that they would be the first targets of any revolutionary uprising, for they represented the wealth and power and privilege of the occupying power. They were the 'capitalist bloodsuckers', as the Mufti of Jerusalem called them, who were bleeding away Arabia's life and strength. In any revolt, it was almost inevitable that the oil installations would be seized and taken over, as a symbol of Arabian possession of its own riches. That the flow of oil would then be interrupted would not matter; it was, in any case, being sent away to feed the British war machine, and why should Arabia's oil be used in a war that was not their war, and paid for in a currency which might soon be worthless?

It was in Kuwait that the British faced their first confrontation with Arab nationalism. The group of local freedom fighters, the Followers, misreading a message passed on to them by Dr Grobba,† took the Bahrain raid as a signal and rose up in revolt. They quickly overcame the local levies and occupied the Kuwait arsenal. Since, thanks to Anglo-Iranian's procrastinations, there was no oil industry to seize they contented themselves by hoisting national flags over the installations at Burgan and taking over the Kuwait Oil Company's office in the town. When the political agent, Colonel Galloway, rushed to see Sheikh Ahmad and asked him to order the rebels to hand over their arms and surrender, he found the ruler exasperatingly dilatory. Why should he not be? He had no reason, no reason at all, to be beholden to the British, who had failed to provide him with an oil income. He had nothing to fear from the rebels, even though they were loudly proclaiming that they were for his removal and for a union with Iraq. They did not know that he had already been in touch with both the Germans and the Italians, and had assurances from both that his powers would be confirmed once victory was theirs.

As the Followers held on to the arsenal and defied all

* He promised his pro-German Iraqi friends that he would be back by 10 May 1941, and was only four days late.

† He had told them to strike 'when the air raids begin'. He meant air raids on Iraq. No one had told him about Italy's raid on Bahrain.

attempts to get them out (they were waiting for the Germans to arrive – six months too early), panic spread through the European community of Kuwait. There were curfews and shots in the night, and urgent plans for evacuating the women and children. Then King Ibn Saud, disturbed by these alarming events so close to his own oil installations, ordered troops to his northern frontiers with Kuwait and threatened to occupy the no man's land (the so-called Neutral Zone) between the two countries. Since Sheikh Ahmad was anxious not to lose any rights in this area, which was potentially rich in oil, he stirred from his lethargy. The rebels were given an ultimatum, death if they defied it, a free passage to Iraq if they gave in. They accepted the terms and departed for the nearby Iraqi city of Basra.

The most serious threat to the whole of Britain's set-up in the Middle East occurred during World War II, however, in Iraq in May 1941, when the military group known as the Golden Square, anti-British, pro-German, but even more fervidly nationalistic, organized a coup d'état and established its leader, Rashid Ali al Gailani, as Iraqi premier in place of a complaisant, pro-British puppet named Nuri Said. Rashid Ali was portrayed in the British and pro-British press at the time as an arch-traitor who used Britain's misfortunes – the British were being hard-pressed in the Mediterranean – as an opportunity to stab her in the back, and that reputation has been written into history. In fact he was no more 'wicked' or 'treacherous' than any other national leader striving to get an occupying power off his country's back, and all he can be faulted for (aside from his failure to bring it off) was his naive belief that the revolt he inspired would bring Iraqi independence with it; for it was obvious that had he succeeded he would merely have exchanged a British collar for a Nazi German yoke.

He had long since made his plans with Doctor Grobba before the Nazi envoy had departed for neutral Turkey, and as soon as he came to power, Rashid Ali summoned the local manager of IPC and told him to assemble two large dumps of motor spirit (about a million gallons in all) in four-gallon tins at Iraqi Army Headquarters in Baghdad and at a post ten miles to the west. By the terms of its contract, IPC was forced to obey, though everyone knew that the dumps were being established ready for the arrival of the German Army. At the same time the new premier reopened the telephone line with Turkey, which had been cut, so that he could talk with Grobba direct, and he re-established contact with Syria, which had a common border with Iraq and had been under the control of the pro-German Vichy regime since the fall of France. Syria would be used as a staging post for the German Luftwaffe's assault on the British in Iraq once the moment came.

That moment was, in fact, timed for 10 May 1941, but the British forced Rashid Ali to jump the gun. They announced that a brigade of the Indian Army was being sent to Iraq to beef up British-controlled forces there (the Iraqi Army was under Rashid Ali's command). Knowing that this would enable the British to quell any Iraqi rising, the Iraqi dictator had to move fast. In any case, he felt that he could hardly lose, for the British elsewhere in the Mediterranean and the Middle East were on the run. They had been driven out of Greece and Crete. Rommel was advancing across the Western Desert towards Egypt. It was just a matter of time.

First he made what seemed to the British ambassador, Sir Kinahan Cornwallis, to be a 'very gentlemanly gesture'. Tensions were mounting and the Iraqi people were getting nervous, Rashid Ali pointed out, and there could be awkward incidents in Baghdad, the Iraqi capital. So why not get British women and children out of the way, so that discussions could proceed without any anxiety as to their safety? The Iraqi leader suggested that they should be taken, under safe escort, to the training camp, run by the Royal Air Force, at Habbaniyah, fifty miles west of Baghdad, and it was thence that they were accordingly sent. Almost immediately Rashid Ali ordered his army to surround the encampment, and let it be known that he proposed to hold them to ransom. If the British gave in, he would spare the women and children. If they did not, he would open fire on them – and do it, moreover, from point-blank range, for the Iraqi Army was positioned in a ring of sandhills only a quarter to a half-mile away from the perimeters of the RAF camp. In the meantime, while awaiting the British ambassador's answer to this nasty ultimatum, he telephoned his friend Grobba and told him that he and the German airforce should get a move on.

It was a grim situation, and no one realized it more than the British oilmen, who were soon feeling the hot breath of Rashid

Ali's minions on their necks. The man in charge of the IPC refinery at Kanaqin was a portly and genial Yorkshireman named Wellington Dix, who was awakened one morning in early May by the barrel of an Iraqi officer's revolver digging into his throat. He was told that henceforward the plant was under the Iraqi Army's control and he and his staff would continue to operate it for the government. He picked up the telephone to speak to head office in Baghdad and had it smashed out of his hand; it would not have been any use anyway, for the line had been cut. He went out into the yard and noticed that every soldier in sight was smoking cigarettes. He immediately ordered the refinery closed down on the grounds of safety, and flatly refused all orders to start it up again.

He and his British staff were promptly arrested and transported first to the jail at Kanaqin and then, after a few hours' journey by train, to the town of Baquba, where they were shut up in the main rooms of the town's principal coffee shop. Their worst sufferings for the next seven days were the blaring of a propaganda radio, the blazing sweaty heat of the rooms, and the menacing shouts of a crowd in the street outside. Otherwise the Iraqis were, as Wellington Dix afterwards put it, 'frightfully British'. They served tea and cigarettes, and were even polite when uttering threats, such as the one made by a fierce-looking army major: 'I am looking forward to the privilege of personally cutting your throat, sir,' he said one evening to Dix.

Three hundred miles to the south-east, John Grafton, in charge of the Rafidain refinery in Basra, was playing a dangerous game. He had the opportunity once trouble began of slipping across the frontier with his staff to Abadan, in neutral Iran, but had refused because a) he could not take all his Indian staff with him, and he was unwilling to let them face the situation alone, and b) because the Royal Air Force had a small station up the railway line at Maqil, and if he closed down the refinery it would mean that their planes would be cut off from aviation spirit; at the same time the British reinforcement's cars and lorries, whose imminent arrival was now signalled, would be deprived of fuel.

Grafton's refinery was taken over by the Iraqi Army and he immediately obeyed the order from the Iraqi commander to put all aviation spirit aboard a series of special trains and send it north to the Rashid Ali authorities in Baghdad. The tins of fuel were loaded into cars, the waybills made out and checked, and the trains duly departed. But they had to pass through Maqil on their way north, and there, warned in advance by Grafton, the cars were unloaded by the RAF and the trains sent on their way.

'When complaints were made,' Grafton said later, 'I blandly produced the waybills which the Iraqis themselves had receipted.'[1]

These, however, were minor setbacks for the Iraqis. The major one occurred at Habbaniyah, where, as Rashid Ali himself remarked later, the British reacted 'in a fashion that was unfortunately typically British'.[2]

To the Iraqi Army which surrounded it, Habbaniyah seemed indefensible, and all its inhabitants, including the evacuated British women and children, in their power. They had brought up their artillery so that it had the landing ground of the encampment under point-blank range from a quarter of a mile away, making it, as the author discovered when he landed there at the height of the siege, a highly dangerous place on which to come down – a shell blew away the tail of the plane in which he landed. Except for rifles, Habbaniyah had no offensive weapons, for all the RAF planes on the station were small mono- or biplane training models, and, in any case, their airfield was too hazardous to use. There were hours of uneasy calm while Rashid Ali waited for news of his German allies, and for his ultimatum to the British to expire. Suddenly made impatient and nervous by news that British troops were arriving in Basra, he ordered his commander at Habbaniyah to give the British a taste of what might be coming to them. Accordingly, the Iraqi gunners put a shell through the screen of the open-air cinema and one through the steeple of the Anglican church, and went on to shell targets sporadically all through the day.

'What I had not reckoned with,' said Rashid Ali later, 'was the British art of improvisation.'

The improvisation consisted of adapting the camp's polo field as an airstrip, loading the training planes with 28-pound bombs, and sending them over to plaster the Iraqis in the sand dunes with shrapnel on as many as 28 sorties per plane per day. While this was going on, and the Iraqi artillery was distracted, a small passenger plane piloted by a Hindu of Indian Airways managed to put down on the main strip. He had flown in from Karachi and he brought in a load of cannon balls, which were promptly fed into the four polished Indian Mutiny cannons outside air headquarters and fired at the enemy, causing little damage but much panic. The plane took off on its return journey with British women and children. A regular ferry service followed and took all of them out of harm's way.

Even so, the revolution might still have succeeded. Though Rashid Ali did not know it, a British rescue column trying to get through from Transjordan had been halted by a mutiny among its Jordanian levies. Indian troops were landing in Basra, but they were many hundreds of miles away from Baghdad.[3] If only the Germans had kept to their promised schedule, everything might still have worked out. Unfortunately for the Iraqis, however, the forces with which the Nazis hoped to take over the country had suffered heavy casualties in their conquest of Crete, particularly in airborne troops. They came late – and not in force. From her refuge in the British Embassy, Freya Stark saw a Junkers 88 passenger plane put down on the airfield and a flight of Messerschmitts sweep over Baghdad, to be greeted by a burst of enthusiastic gunfire from the Iraqi troops on the field. It was 14 May 1941, and from the Junkers stepped Doctor Fritz Grobba, the Nazi envoy, four days late for his promised appointment with the rebels. Rashid Ali and his cabinet were there to meet him, the idea being that there would be a formal banquet at which Iraqi-German unity would be announced and the Nazi army invited to take over the campaign against the British.

It did not happen in quite that way. The banquet was cancelled when it was discovered that one of the Messerschmitts, piloted by Colonel von Blomberg, who was to take over operations, had crash-landed. One of the bullets fired by a rejoicing Iraqi soldier had pierced von Blomberg through the eye and killed him instantly. It then became obvious that Dr Grobba was by no means clear about the military situation, and thought the whole country was in Rashid Ali's hands. When he discovered that this was by no means the case, he hastily flew back to Syria en route for Turkey.

Wellington Dix of IPC knew it was all over when his captors transferred him and his companions from their foetid quarters to an air-conditioned room, and offered a meal of curried bully beef and Australian beer. Grafton knew it when the provincial council of embarrassed Iraqis came to the office in Basra where he was confined and announced that he had been unanimously named military governor, and would he please henceforward act as an official link with the British forces.

There had been some damage to IPC's properties and rough handling of its Indian personnel, and one river tanker was sunk when its crew tried to make for the sea. But otherwise damage was light, and a report published shortly afterwards by IPC recognized that most Iraqi personnel went on working during the rebellion 'and it was largely due to their efforts . . . and, it must be admitted, to the reasonable attitude adopted by the [Petroleum] Ministry's officials that no dislocation of the company's activities and records occurred.'

To which Rashid Ali said later: 'Of course there was no interference. That was by my express orders. Why should we damage property which was ours, and which, one day, will come into our rightful control?'

That, however, was still a long way away. At the end of May 1941, as the British took back control of Iraq, Rashid Ali and forty of his followers (naturally dubbed 'Ali and the Forty Thieves' by the British) fled to Iran, and the first major Iraqi rebellion was over.

In June 1941, the Germans launched their attack upon Russia, and Winston Churchill's promise that he would help to keep the USSR supplied with any materials she needed gave Britain the opportunity to straighten out what Whitehall called 'some little local difficulties' in Iran. For there were only two ways of getting supplies through to the Soviet Union: by sea from Britain along a route harassed every murderous mile by Nazi submarines or planes, or from the Middle East by way of Iran and through to the Russian Caucasus.

But Iran was neutral, and, as Anglo-Iranian had been finding recently, irritatingly determined to make that clear beyond all doubt. Customs officials who had hitherto been liberal in their interpretation of regulations were suddenly meticulously and painstakingly following the complicated letters of Iranian import laws. Red crosses on ambulances and on the hospital at Abadan refineries were ordered to be painted out. Tankers arriving from the war zones had to remove the breech-blocks of their guns. The shah-in-shah's finance minister demanded from the company an immediate loan, free of interest, of $9,000,000 and a guarantee, whether it was earned or not, of $16,000,000 in oil royalty payments.

There was little doubt that the shah had decided that Germany was going to win the war, and it was a prospect which did not displease him. He hated his Russian neighbour to the north, with good reason, for Iran had long been the victim of Russian depredations and threats. Except that they used friendlier tones when making their demands, and that their representatives were rather more personable, he regarded the British as little better than the Communists. When negotiating with Iran, the British also always made it clear that they had a gunboat up their sleeve.

To give himself a counter-balance against Russian political power to the north and British economic power in the south (through its control of the oilfields), the shah-in-shah had begun calling in German experts to help him even before World War II began. He was anxious for the independent industrialization of his country, and he knew of no better technicians than the Germans to help him towards that goal. From his point of view, it was a welcome bonus when it turned out – by 1941 – that they also belonged to a nation on the point of conquering both Russia and Great Britain. By that time there were around 3,200 German 'experts' of one kind and another in Iran, and though most of them were indeed genuine engineers, architects, scientists and surveyors, it was inevitable that among them was a hard core of Nazi agents and shock troops. The fact that they were sitting on the threshold of Britain's lifeline to India and the Far East, and that they were moving around freely in the area where she was operating her biggest oilfield, caused quite a few shudders in Downing Street. But until Germany attacked Russia, there was little that Britain could do about it. The shah was convinced that Germany was going to win the war. The Germans were popular with the more influential Iranians. British remonstrances only made the shah more determined, the Iranian Army more hostile, and the gendarmerie in Abadan and other Anglo-Iranian oil centres more rigidly unpleasant in their controls.

The shah's great mistake was to persist in his unswerving

and overt pro-German policy even after Russia was attacked. So sure was he of an imminent Nazi victory that he waved them aside when the Russians added their protests to those of the British about the German 'fifth column' he was harbouring. To underline his contempt for Britain, he immediately granted sanctuary to the Iraqi rebels when they came over the frontier into Iran after the collapse of their rebellion, and he publicly welcomed their leader, Rashid Ali. It was the last aggravation. On 25 August 1941, after secret consultations with the Kremlin, British and Russian troops crossed into Iran and occupied the country, The Russians came in from the Caucasus and took Azerbaijan. The British made for the oilfields in the south and took no chances. In the half-light of dawn the Royal Navy cruiser *Shoreham* crept up the Shatt al Arab river towards the refineries at Abadan and the city of Khorramshahr, just across the river. The Iranian sloop *Palang* was lying alongside number eleven jetty, and the British commander opened fire on her without warning, in case she might be tempted to shell the tank-farm and refineries behind her. In five minutes the Iranian war vessel was a blazing wreck. In similar and salutary fashion, HMS *Yarra* was sending another Iranian sloop, *Babr*, to the bottom off Khorramshahr.

But then things went wrong for the British. A landing party sent to take over the refinery went aground on a mudbank, having misread the river channels. By the time a second wave of Indian troops and their British officers were ready to go in, the Iranian Army had reacted. They had been given no orders from Teheran, so they decided to fight. The second wave of Indian troops, not having been briefed for first-wave operations, got into difficulties struggling over barges lying between themselves and the refinery waterfront, and came under murderous machine-gun fire from the Iranians. Almost immediately they lost their colonel and two other British officers. It took them some time before they managed to clear the dockside and force the Iranian garrison back into the general office of the Anglo-Iranian Company, a building outside the iron rail surrounding the refinery.

It was the first time these particular Indian troops had ever been in action, and now they had lost three of their British officers in the first fifteen minutes. They had been told at their briefing that they were going into Iran to clear the country of

Germans, and believed they would be fighting the Nazis. What they were looking for was any European or anyone wearing a uniform—these were the enemy, they believed. So when they came into the main entrance of the general office building and saw a group of men in blue uniforms, they opened fire at once – and slew eight messengers in the blue livery of the Anglo-Iranian Company. One of the company's general staff, George Wheeler, saw the slaughter and at once telephoned the company hospital for an ambulance. It was ambushed by the Indians as it came whizzing to a halt, for the red crosses on its sides had been painted out on the shah's orders. The driver and attendants were killed.

Wheeler and two of his colleagues, now coming under fire from both Indians and Iranians, took refuge in a corridor and were mown down by an Indian machine-gunner. Wheeler, badly wounded, managed to shout in Hindustani that he was British, but by then it was too late. His two companions were dead. But he did manage to convey to the Indians that their targets were Iranians, not Europeans, and by noon they had the refinery under control. The main body of the Iranian Army now began to retreat across the river, and set off across the blazing desert towards the city of Ahwaz. They took four British oil company representatives with them. It was a ghastly journey and fifty-seven Iranian troops died of thirst en route. The British survived because the Iranian soldiers shared their rations of water among them.

At Masjid-i-Sulaiman, where Anglo-Iranian's fortunes had begun, the British staff persuaded the local Iranian Army garrison that to stop work would mean that 'if a well got out of control, the first persons to be gassed would be the police and troops on the Maidan and in the bazaar. Furthermore, were the pumping station to be shut down, everyone, beleaguers and beleaguered, would perish of thirst. These representations had the necessary effect and no harm was caused to the wells or the pumps.'[4]

It was even more frightfully British out in the fields around the Iranian-Kurdish city of Kermanshah. The company's local manager, A. W. M. Robertson, an amiable Scot, was personally informed of the Anglo-Russian occupation by the Iranian commanding officer, General Hassan Muqaddam. They discussed over a drink how best to protect the refinery 'from possible looting by Kurds in search of food, for these tribesmen were on the edge of starvation.'

Work went on as usual and each morning the Iranian general reported the progress of the war to the manager. On Thursday 28 August, which is the normal half holiday, the British staff assembled at the Club and arranged to play a fourball match on the golf course which they had built. They set out but without Robertson, who had been hastily sent for by General Muqaddam and asked to take himself and his staff away from Kermanshah at once. It transpired that the Shah had telegraphed an order [to the general] to surrender, but the British commander, unaware of this, and fearing for the safety of the Britons of the oil company, had also warned him that unless the staff were assembled intact at a spot outside Kermanshah by 2.30 the garrison in Kermanshah would be bombed.[5]

Robertson told the agitated Iranian commander that it was not possible for the staff to leave Kermanshah at the moment.

He explained that he could not break up a game of golf but would go himself. This he duly did, setting out with the general's chief of staff and a white flag flying from the bonnet of his car.[6]

Later, at a meeting between the Iranian general and Major General (later Field Marshal) W. J. Slim, the British commander of the operation, Slim remarked: 'The war between us would never have occurred if the Shah had permitted us to come in and take the German fifth columnists.'

But they had gone underground or crossed the frontier, as Rashid Ali and his friends had already done, into Turkey. But that had not really been the object of the exercise. What the British and the Russians wanted was not a genuinely neutral Iran, as they claimed, but a definitely pro-Allied one. That they now had. Rezah Shah, resentful and embittered, was removed from his throne and the crown handed on to his son, the present monarch. The old man was taken into a guarded exile first in Mauritius and later in South Africa, where he died, his enemies unforgiven, on 26 July 1944.

His ashes were not returned to his own country until 1950, and when they came, so far as Britain's oil stake in Iran was concerned, the chickens also came home to roost.

11 · Shut-Down in Arabia

When Italian planes carried out their surprise attack upon Bahrain in October 1940 they also dropped a spray of light bombs on the American oilfield at Dhahran in Saudi Arabia, no doubt by mistake because, of course, Saudi Arabia was not involved in the war and the United States was still neutral. The bombs made a lot of noise but fortunately did almost no damage, but there was, nevertheless, a high degree of alarm in the Aramco camp while they were exploding. It was not very long since Dhahran had come through the most alarming moment of its existence, and that was when one of the wells had caught fire. Everyone shuddered at the thought that it might be happening again.

Fire is the worst thing that can happen in an oilfield. It can blow everything to bits on the surface around it. It can cause havoc to the reservoir of oil underneath the ground. It is hellishly difficult to put out.

It was in the stupefying heat of a July afternoon in 1939, at a moment when Dhahran was cabling home office: 'Good news. Dammam Number Twelve is on its way,' that Dammam Number Twelve exploded. No one ever worked out why. The well had been test-probed to 4,565 feet and oil found in promising quantities, and now a crew was in place ready to take the bit down a further stage into the petroliferous strata. No one expected any trouble. The driller, Bill Eisler, had his perforating gun ready to make the downward stab; a Saudi workman was down in the 'cellar' under the floor of the derrick with his hand ready to open the equalizer valve; three other crew stood by the main valve controls; and Monte Hawkins, a second American driller, was by the drill-hoist.

Suddenly shock waves rolled like great invisible breakers through the heavy air, followed by an enormous low grumbling thunder of sound that seemed to feed on itself and grow and grow. Dammam Number Twelve spouted yellow and purple flames which leaped, in an instant, three hundred feet into the air. Hawkins, knocked flat, turned and saw Eisler falling off the derrick, surrounded by flames, scrambling to try to get clear. He rushed forward to grab Eisler by his cindered hands and somehow, with help from others, got him away to a car, but he died not long afterwards in hospital. The Saudi workman in the 'cellar' had somehow managed to scramble out, just in time to miss being felled by the derrick, crashing to the ground as its steel girders melted at the roots. The fire was now roaring away, wailing like a banshee, a red twister of flame so hot that it blistered the paint off trucks parked two hundred yards away.

Soon everybody in the oil world knew that Dhahran had a super conflagration on its hands. Experts from the US sister company in Bahrain announced that volunteers were on their way to help. From Abadan Anglo-Iranian cabled that they were dispatching a team of fire specialists. Charley Potter, the company's drilling superintendent, on leave in the US, rushed to New York to confer with Myron Kinley, the Texan with the reputation of being the world's greatest oilfire fighter. He declared himself ready to come over with his team on a chartered plane, a much more difficult journey in those days than it would be now.

But all this would take time and fire-fighters were needed now, and special equipment: gas masks, asbestos screens, fresh-air suits, extra firehose. Aramco's London office was alerted to go on a buying spree and fill a chartered plane with the desperately-needed supplies. Britain was on the eve of World War II, expecting conflagrations of its own, and refused at first to release the equipment. When it was at last put aboard a chartered plane this was only allowed to fly as far as Rome, where it was off-loaded on to an Italian plane. This in turn was grounded by the British in Basra (determined that no Italians should fly over their oil installations) and the equipment brought on in an RAF plane to Bahrain and thence to Dhahran. Meanwhile, the fire roared on.

Because No. 12 was a good distance from the other wells [writes Wallace Stegner], there was no serious danger of the fire's spreading.

The real danger was that the master valve and the connections on the main casing would be destroyed, which would probably destroy the well, and might also spray the entire camp with burning oil. Besides, if the well ran wild, it might seriously deplete the whole oil field by releasing gas pressure and possibly chanelling water into the oil zone.[1]

It took ten days to bring Dammam Number Twelve under control and douse the whirling fire. Not only did the world know about it now, but also King Ibn Saud, and he, convinced that his oil wealth was going up in flames and smoke, demanded hourly news of its progress and talked pessimistically about the wrath of God. One particularly fervent religious mullah announced that he proposed to immolate himself upon the pyre in the hope of appeasing the malevolent furnace, but only if his sacrifice was followed by a purge of the evil and impious who were gaining influence in Saudi Arabia, and by the expulsion of the infidels who were encouraging them. No doubt by arrangement, his friends dissuaded him at the last moment, but he was to be found among the crowd of gawping locals, hectoring them with many gestures and warning them to repent.

Meanwhile, the fight to kill the fire went on. It took long conferences and much preparation even to plan how to deal with it. The men who were fighting it were in no sense professional fire-fighters, but it looked as if the professional fire-fighters would never get there before the well blew up completely, and so they had to improvise, something which, luckily, oilmen excel at, and they were fed with advice by cable from Charley Potter and Myron Kinley in New York, and by oil offices all over the world. Finally they had asbestos screens up, enclosing the wellhead, and had found enough water – which had been scarce to begin with – in a salt-water well to give them all they needed for the hoses. Three men in asbestos suits volunteered to go to the head of the well and see what they could do there.* They had studied the plans and knew there were two large gate-panels over the well, and if they could get them closed down it might help. Sprayed with water from the hoses to help keep them cool, they went through a

* The volunteers were Bill Eltiste, Herb Fritzie and Walt Sims, all Aramco drillers or ex-drillers.

gap in the asbestos screens and by extraordinary tenacity ('it scorched your eyeballs,' Eltiste said) they managed to reach the wreck of the wellhead and for thirty seconds tried to get the control wheels turning. One went round twice and then stuck. The other wouldn't budge at all. Next day they went in again, and before they were dragged clear they managed to get one panel half-down. It slightly diminished the flame, and enabled Eltiste to see where the fire was coming from – through a split nipple in the control valve.

It was then that home office in San Francisco came up with a suggestion: Don't go for the well direct; dig a tunnel underground to it, put a tapping machine on the base of the main valve, and start pumping mud into the hole to smother the flaming oil. That was easier advised than carried out, but on 18 July, ten days after the first mighty explosion, they had a tunnel built and, thanks to the half-closed panel-gate, had managed to get a 'hot tap' ready to slip on to the valve. Two men,* working night and day in a great heat and peril, hanging behind shields, head first into the cellar of the well, had moved the connecting line into place. Now all that remained was for the bit to drive a hole in the main valve and for the mud to be pumped in before the flames from the well could spurt out and through the tunnel. They played it by feel rather than by ear (though the fire was only a subdued roar now) and when they felt the bit judder through Eltiste raised his hand and the pump began gushing and slopping gallons of thick mud into the fiery hole. Like a damp candle, the flame just guttered away.

'FIRE EXTINGUISHED. HOLE FULL OF MUD. PROFESSIONAL FIRE-FIGHTERS NOT NEEDED,' cabled Floyd Ohliger, the general manager, to New York, where Myron Kinley was still waiting for a plane.

There was no expression of pride and joy in the message, but it was implicit between the lines.

In celebration of the way they had handled themselves [wrote the official historian], the [Saudi] Government—for this one occasion—relaxed its prohibition law, and the first beer that was ever in Al Hasa came across from Bahrain, and the firemen really tied one on.[2]

The great Dhahran fire cost the lives of an American and a Saudi, did $3,000,000 worth of damage, and deprived King Ibn

---

* Cal Ross and Ed Braun, two more drillers.

Saud of half a million dollars in oil royalties. But more important, it made the men who ran the oilfield feel vulnerable, and
scared. When the Italian bombs dropped upon them (probably,
as has been mentioned, by mistake) it is not too much to say
that they panicked. America was not in the war, nor was Saudi
Arabia, and there was no reason at all why the Axis powers
should breach the neutrality of a kingdom whose ruler both
Germany and Italy were anxious to placate. The immediate
measures which were taken after the raid were, perhaps, well
justified, though messy or complicated: roads and buildings,
for instance, were painted with oil to take the sheen off them,
and the gas flares which flicker over every oilfield were strained
and treated so that they burned a dim blue instead of a fiery red,
to make them less of a give-away from the air. But more
drastic was the decision, taken after a pregnant wife got
frightened and hysterical, to evacuate all American women and
children, plus all male employees who wished to go. At the
beginning of 1940 Aramco's US complement had been 371
employees, 38 wives and 16 children, and a year later it was
down to 180 males and no women and children at all. Worse
than that, the whole operation was slowly running down.

Thanks to Max Steineke and a dogged fellow geologist named
Tom Barger (later a vice-president of Aramco), two new fields
pregnant with possibilities had been discovered in Saudi Arabia.
One of them, Abu Hadriya, deep into the Rub al Khali (the
Empty Quarter), had struck oil at 10,115 feet in March 1940,
and just about the time of the bombing, in October of the same
year, a well had gone down at Abqaiq, about forty miles into
the desert from Dhahran. Both of them promised enormous
potential* and everyone was bubbling with optimism over
their prospects. But immediately after the raid, Abu Hadriya
was ordered closed down and the following spring Abqaiq
followed suit. Thanks to the evacuation of personnel, there was
not the manpower to handle them. By the beginning of 1941
only Dammam was operating, and that only at the rate of
about 15,000 barrels a day, a mere trickle compared with the
field's potential. The crude was ferried by barge across to the
refinery at Bahrain, and Aramco's own brand new installations
at Ras Tanura, its huge tank-farm, its small refinery, its port

* Abqaiq has since become one of the most prolific fields in the world.

facilities for tankers, lapsed into disuse. A gloomy hush fell over what, a few months before, had seemed like the most promising new oilfield in the Persian Gulf. And as word spread that the Aramco operation had closed down for the duration, the benevolent vultures in other parts of the Middle East and India came around looking for pickings. Especially after Pearl Harbor brought the United States into the war, theatre commanders all over the area began calling on Dhahran for loans of trucks, pickups, welding equipment and drills, sapping the supplies which Floyd Ohliger had painstakingly built up in 1939 against the shortages of war. Now they could not use them, and had to give them up. And soon they were so pinched for transport that they organized a camel corps to transport packages around the fields.

No one was more distressed by the close-down of Aramco than King Ibn Saud, for it meant a drastic cut in his income, a blow to his hopes for its increase in the future, and it could not have come at a worse time. Other than oil royalties, his only revenues came from money brought in by pilgrims to Mecca. In a normal year more than a million* devout Muslims came to Saudi Arabia for the *haj*, but now nearly all the countries from which they came (the Middle East, India, the Far East and Africa) were involved in the war, and no ships were making the journey. To add to his distress, the winter rains had failed and there was a drought in the Al Hasa oasis, killing the crops; the wells were dry and the bedouin were eating their camels before the vultures got them.

The king needed money badly, to pay off his old debts, of which there were always plenty, to relieve his subjects' distress, and to pay for his own extravagances. Perhaps extravagances is not the right word, for though all the money the state received was paid, in the first instance, directly to the king, what happened to it afterwards was left to his trusted ministers – who, in fact, could not be trusted. Ibn Saud asked for money and was always given it, to pay for new wives or concubines, to subsidize tribes which might otherwise have become dissident, to satisfy the importunate demands of his myriad sons and their acquisitive appendages. But the purse-strings were in fact controlled by his finance minister, Abdullah

* Nowadays as many as 2½ million make the pilgrimage.

Sulaiman, and if he kept a strict accounting of what was done with all the monies flowing into the king's coffers, no one else saw it. Ibn Saud was too bored to ask about it; he was not interested in accounts; all he wanted was money to pay his bills.

On 18 January 1941 Abdullah Sulaiman, writing on behalf of the king, sent a formal letter to Aramco noting the harsh facts of the financial situation in Saudi Arabia and requesting the immediate payment ('as an advancement on oil royalties') of $6,000,000. He asked the company to take note of the fact that he was reserving the right, while making the present request, of continuing to do so for further advance sums until such time as the emergency in Saudi Arabia no longer existed. He also underlined the fact that the Americans had recently been granted extension of rights and concessions, and that these had shown unlimited possibilities for successful exploitation. F. A. Davies, the director in Saudi Arabia, telegraphed the minister in Jiddah that he had already arranged for $3,000,000 to be advanced to the Saudi government, but that at the same time he had been instructed to inform the king that 'as long as the war continues and the market for Arabian oil continues curtailed we do not want to increase amount advanced nor to lead him to feel that this is precedent for further advances in the future.' As a mollifying gesture he described the mission he and a fellow director had made to San Francisco in an attempt to persuade the board to increase the subvention to the Saudis, and he added:

In our many discussions with His Majesty and Your Excellency we have come to a full realization of the Government's need for extraordinary assistance during the present international emergency. Should this emergency extend beyond the year 1941, the Company will, of course, continue to assist as much as it can . . . Your Excellency need have no doubt that the Company realizes fully how closely its own interests are bound up with those of His Majesty's Government.[3]

Nonetheless, the truth had to be faced. The glorious future of Saudi Arabian oil had turned rancid, and at the moment the company was $34,000,000 down on its investment. It was not prepared to pay out any more.

On the other hand, the US government might be persuaded to do so. It was known that they were pleased at the inroads the

Americans had made in Saudi Arabia, an area previously jealously guarded as a British preserve, and perhaps the State Department could be persuaded that there was a danger of losing the advantage so handily gained.

It so happened that one of the directors of Aramco was James A. Moffett, who had once been a member of the State Department himself and counted the US president, Franklin D. Roosevelt, among his friends. At a moment when Abdullah Sulaiman was at his most importunate for a $6,000,000 loan, Moffett went to the White House and had a discussion with the President in which he outlined the king's difficulties and the company's situation, at the same time stressing the strategic importance to the United States of the Saudi oil concession. He asked for a US government loan to Ibn Saud to tide him over. Roosevelt replied that there was no way in which such a loan could possibly be made under existing American laws, but, as an ex-secretary of the navy, he suggested that perhaps something might be arranged by selling oil products to the US fleet, and told Moffett to go away and write a memorandum to the government explaining the company's present plight and that of the king, and outlining a plan whereby a five year oil sale might be made to the navy that would provide the king with the $6,000,000 he so sorely needed.

In the years to come, US senators were to accuse US oil interests in Arabia of being a government unto themselves; and in parts of his memorandum Moffett implicitly admits the supranational nature of the company's commitment.

It has now come to a point [the memo stated], where it is impossible for the company to continue the growing burden and responsibility of financing an independent country, particularly under present abnormal conditions. However, the King is desperate. He has told us that unless necessary financial assistance is immediately forthcoming he has grave fears for the financial stability of his country.[4]

But now Moffett gave the impression that his company was ready to abdicate its responsibilities and, if the US government refused to accept them, pass them on to the British. In fact neither he nor his company were prepared to do any such thing, but they had 168,000 US shareholders to think of and enormous subventions could not continue to be paid out of the oil company's funds at a time when its Saudi Arabian investment was stagnating. So what happened next was a shrewd ploy to relieve its own resources of the Saudi burden and get the US government involved in carrying it instead, without appearing to have done so.

Moffett duly outlined in his memorandum a plan for a US naval purchase of Saudi oil, as President Roosevelt had suggested, but he had little hope that this was the way to find the money Ibn Saud was demanding. His suspicions were confirmed when he had an interview with Secretary of the Navy Knox who told him at once that the amount of oil that would be involved in a $6,000,000 programme was far beyond the needs of US navy ships in the Persian Gulf and the Indian Ocean. When Moffett expressed the view that the navy could buy the oil as a future reserve, the navy secretary replied that though it might be a sound idea he had no funds whatsoever for such a purpose.[5]

It was for this reason that the oilman included in his memorandum another way around the problem, one that would mean bringing in the British government. He was well aware of the fact that the British had strong political and strategic reasons for being concerned over the future of Saudi Arabia. So far as the war was concerned in the Middle East, they were in a bad way, with General Rommel and his German Afrikakorps waiting in the Western Desert and knocking on the gates of Cairo and Alexandria. British backs were to the wall, and the wall was Saudi Arabia, and therefore far from stable. There were already strong indications that certain influences in Ibn Saud's court were in the pocket of the Nazi government, while the king himself was reluctantly coming to the conclusion that Britain was losing the war. He did not particularly like the British, but he trusted them more than the Axis powers, especially Italy, whose activities in occupied Ethiopia, on the other side of the Red Sea, made him fearful of what might happen to his own country if it came within the Italian orbit. He was therefore inclined to go on tingeing his neutrality with pro-British hues as long as he possibly could; but in order to do so he must have enough financial resources to alleviate distress in his own country and to prevent his people from being bribed or browbeaten by Axis propagandists. To bolster him against enemy pressures, and to make up for the loss in oil revenues from the Americans, the British government had already begun loaning the king an annual sum of £400,000 ($1,600,000 approximately at the then rate of exchange). Moffett urged the State Department to persuade the British to increase their annual contribution to Ibn Saud's exchequer, and to make loans of its own to such an amount that, together with what small revenues the king did receive from oil and pilgrims, would provide him with the $10,000,000 a year the oilmen estimated he needed to keep Saudi Arabia solvent and peaceful.

Moffett added one point to his memorandum which made it quite clear that his company, though dormant, was far from being willing to see its proprietary rights in Saudi Arabia taken over by anyone else. He asked the State Department to make it quite clear to the British that in making their loans to the king they were doing so for reasons of war strategy only, and their aid would in no way give them any claim over the oil concession.

It was a moment, fortuitously enough, when the federal loan administrator, Jesse Jones, was in the throes of working out a Lend-Lease loan to Britain of $400,000,000 and after a discussion with Harry Hopkins, the presidential adviser, and Secretary Knox, President Roosevelt suggested that the best solution to Saudi Arabia's problem might well be for the British to apportion part of this sum to the king. On 12 June 1941 he sent a note to Jones saying:

Jesse: Will you tell the British I hope they can take care of the King of Saudi Arabia. This is a little far afield for us! F.D.R.[6]

At the same time Harry Hopkins wrote a letter going into the matter from a slightly different angle, wondering aloud whether Saudi Arabia could itself qualify for the type of Lend-Lease loan which Congress had agreed to supply to the United States' 'democratic allies'.

Dear Jesse [he wrote on 14 June 1941], The President is anxious to find a way to do something about this matter. I am enclosing confidential correspondence from the White House so you can see what goes on. Will you return it to me as soon as you have read it?

I am not sure what techniques there are to use. It occurred to me that some of it might be done in the shipment of food direct under the Lend-Lease Bill, although just how we would call this outfit a 'Democracy' I don't know. Perhaps instead of using his oil royalties as collateral we could use his royalties on the tips he will get in the future on the pilgrims to Mecca . . . Cordially Yours, Harry.[7]

It was eventually the British who agreed to take over the burden of keeping King Ibn Saud afloat (admittedly out of the US government's Lend-Lease loans) and for a time the American oilmen were full of self-congratulation on the way they had successfully plugged the drain on the company's resources. Their representatives in Jiddah took pains to point out to Abdullah Sulaiman, and on two occasions to the king himself, that they must regard the British subventions as a substitute for their own aid, and not as additions to the loans and royalties the Saudis had been demanding from the company. In other words, the British were bailing them out of a most awkward situation, and saving them many millions of dollars which otherwise would have had to come out of the company's exchequer. How much that saving was is indicated by some figures which were quoted by Floyd W. Ohliger, vice-president of Aramco, at the Senate Inquiry into Petroleum Arrangements with Saudi Arabia at their 1948 hearings in Washington. Ohliger testified that the company paid the king as loans against future royalties the sums of $2,980,988 in 1940, $2,433,222 in 1941, $2,307,023 in 1942, and $79,651 in 1943. It was a downward trend that must have brought joy to the treasurer at home office in San Francisco. On the other hand, over the same period the British government's loans to the king went upwards even more steeply than the oil company's had gone down: $403,000 in 1940, $5,285,500 in 1941, $12,090,000 in 1942, and $16,618,280 in 1943.[8]

Were the British motivated solely by their desire to keep Saudi Arabia in the Allied camp? That might have been true in 1941 and 1942, but by 1943 the danger that the Axis would sweep across the Middle East was over. Italy had been driven out of Ethiopia and Eritrea, the German army out of North Africa. Iraq, Syria and Palestine were securely in British hands. On the other hand, Washington's disinterest in Saudi Arabia and the American oil company's abdication of its declared responsibilities had left a vacuum which London was only too pleased to fill. There is no evidence that a direct attempt ever was made by the British to take over the American concession, but what would have happened if, out of gratitude, King Ibn

Saud had offered it to them – angry as he was with the 'defections' of the US oil company? The British government held a 51 per cent controlling interest in Anglo-Iranian Oil Company, and by a lucky coincidence the director of military intelligence at British GHQ in Cairo was none other than Archibald H. T. Chisholm, once an executive of Anglo-Iranian and negotiator of its Kuwait concession, now a brigadier, his monocle sparkling over his crimson staff officer's tabs. So far as Intelligence operations were concerned, Saudi Arabia was well inside his territory.

Suddenly, Saudi Arabia, which had once seemed 'a little far afield' to Franklin D. Roosevelt, began to seem much nearer as British influence began to make itself felt at King Ibn Saud's court. It was the moment, the oilmen decided, to sound the alarm and bring the US government to the rescue.

'We didn't like it,' testified W.S.S. Rodgers of the Texas Company later.

> We realized that he [King Ibn Saud] had to have the money, but we didn't like the British Government making these advances. . . . And then at the beginning of 1943 it got so bad that we came down here to Washington and called on many people, trying to get the matter straightened out.[9]

Mr Rodgers appears to have ignored the fact that it was his company which suggested that Britain should increase her subventions to the Saudi Arabians, in the first place. He now saw the situation in a very different light. In a memorandum submitted to the Senate Investigating Committee, he described it in this way:

> The situation in Saudi Arabia had now [1943] reached the point that the British were backing the Saudi Arab Government so far as its finances were concerned. It would appear that they were looking for an opportunity to remain as the financial advisor and backer of the Saudi Arab Government without having to advance any actual gold or silver. This crystallized in March 1943 when they proposed a plan for a Saudi Arabian note issue. This plan provided for the creation of a Saudi Arab Currency Control Board in London, composed of the Saudi Arabian Minister, Government of Great Britain Representatives, and Bank of England Representatives.[10]

It was time to call a halt, Rodgers said, in the series of conferences he had in Washington (in February 1943) with members of President Roosevelt's cabinet, and the one upon whom his alarm calls seem to have made the deepest impression was Secretary Harold Ickes. It was Ickes who carried the campaign to the President and convinced him that the British were threatening the US company's oil concession in Saudi Arabia, and that action must be taken before it was too late.

Altogether, the moves had been well timed. There was much discussion going on in Congress and in the newspapers about America's future oil reserves, and the chiefs of staff were making known their desire to have a reserve of petroleum in the Gulf area for the use of the armed forces as the war moved towards the Indian Ocean and the Pacific. On 18 February 1943, President Roosevelt decided to twist the terms of the Lend-Lease Act even more than they had been manipulated so far in the war. In a letter to Edward Stettinius he wrote:

My Dear Mr Stettinius :
For purposes of implementing the authority
conferred upon you as Lease-Lend Administrator
by Executive Order No 8926, dated October 28,
1941*, and in order to enable you to arrange
lend-lease aid to the Government of Saudi Arabia,
I hereby find that the defense of Saudi Arabia is
vital to the defense of the United States.
Sincerely yours,
Franklin D. Roosevelt.[11]

This meant that King Ibn Saud at once became the recipient of the bounty of the United States Treasury, along with America's fighting allies. Once more, Ibn Saud's insatiable needs were appeased and Aramco's concession saved. First the British had come to the rescue; now Washington. And it had not cost the oil company a cent.

For such invaluable service, the British naturally did not expect any gratitude, for their charitable gestures had not exactly been devoid of self-interest. On the other hand, the munificence of the US government† – basically for no other reason than to disembarrass a commercial enterprise and relieve it of a heavy financial obligation – might have been expected to produce some tangible evidence later of the

* Which recognized Soviet Russia as a 'democratic ally' for the purposes of the act.
† It gave Saudi Arabia $99,000,000 under Lend-Lease.

company's gratitude. But when Washington asked the oil company for a gesture in return, it found that the company was, in Harold Ickes' words, 'more disposed to thumb their noses at us'.

Those were the wartime days when the US government had some say in the control of almost every branch of American industry. Through boards established throughout the nation, raw materials were allocated, markets were controlled, prices were fixed, and wasteful competition eliminated. The petroleum industry had its own committee through which production rates of crude, petrol and aviation spirit were decided, and there were some oilmen – though by no means all – who said that the industry worked better that way. The President and members of his cabinet agreed with them, none more so than Harold L. Ickes. He went one stage further. He believed that what would be best for the US government and people would be for the government to buy into the oil industry and secure a controlling interest in a major source of supply, so that Washington would have a say in pricing, production and policy, and would cease to be at the mercy of commercial enterprises interested solely in keeping up prices and profits. What the British government had done to Anglo-Iranian in 1914, and what the Arabian and Iranian governments began manoeuvring to do to all the major oil companies in 1972, Ickes proposed that the US government should do to the American owners of Saudi Arabian oil in 1943. In other words, buy a share of the enterprise.

First of all (in June 1943) he wrote a note to the president proposing that a Petroleum Reserve Corporation be set up to 'acquire and participate' in the exploitation of foreign oil reserves, and he suggested that the first task of the new corporation would be to acquire a 'participation' in the Saudi Arabian concession. It would be a move which would effectively, and once and for all, prevent Britain (or 'certain known activities of a foreign power', as Ickes put it) from taking over. To Franklin D. Roosevelt, who was not known for his love of oilmen (who were apt to vote Republican), it seemed a capital idea, and he forthwith set up the Petroleum Reserve Corporation with Ickes as president and chairman of the board. At once the secretary summoned officials of the California Standard Arabian Oil Company (as the company was still known then) to see him in Washington and announced to them that he proposed to begin negotiations for the purchase of an interest in their enterprise.

A year earlier, when the whole Middle East seemed on the verge of being overrun by the Axis, and when the oilmen were mentally writing off Saudi Arabia, this was a proposal which they might have welcomed as some form of government insurance of their property, and therefore worth making a sacrifice of part of their control. But by this time, the situation was altogether different; there was no question of a military debacle; the government had taken care of their financial obligations to King Ibn Saud; and they could afford to think about postwar production and profits again. So their attitude was hostile from the start. When they asked Ickes what he considered would be the extent of the government's participation in their company, the secretary replied that it would solve a lot of problems for the Petroleum Reserve Corporation if Arabian Oil were to sell its entire stock to the corporation, for an amount to be agreed plus a royalty. Ickes said later that this casual remark so frightened them that 'they nearly fell off their chairs'. The oilmen asked permission to retire to consider what the secretary had said, but in reality to get their breath back and to spread the alarm through the oil industry that the heat was on and that the US government was moving in on them.

Nothing is more calculated to unite a normally competitive and divisive industry than a threat to its future, and as the firebells rang through the oilfields, tycoons from Texas to California and from Oklahoma to Venezuela and the Caribbean began ringing Washington and lobbying their friends in the Senate and Congress. Aware that his casual remark had set the cat among the pigeons, and done it too soon, before the cat was ready to pounce, Ickes called the Arabian oilmen together again and began reducing the amount of participation he thought the US government should obtain. First it was 'whittled down', as he termed it, to 70 per cent, then to 51 per cent. As W. S. S. Rodgers, one of the directors involved in the discussions, later testified:

Naturally that did not appeal to us. Then they got down to a 33⅓ percent basis and we did not like it, or at least I did not personally, but we began to get a little closer. We did not get close enough, and one day Mr Ickes said, I don't know why, 'The negotiations are off', and I was very much relieved.[12]

Mr Rodgers may not have known why Ickes suddenly gave up, but most other people in the oil industry did. Too much pressure had been put on Washington by irate oil interests at a moment when the government needed all the co-operation it could get to ensure a steady supply of fuel for the armed forces in the crucial build-up for the last phases of the war.

They came up here to the Hill and built a fire under us on the theory that this was an attempt on the part of the Government to take over a private-business enterprise [Ickes testified later], which, of course, was against the American tradition, as they put it, and perhaps it was. But this was more than a business enterprise, this involved the defense and safety of the country.[13]

He had expected that at least the American oilmen operating in Saudi Arabia would be willing to admit some measure of US government participation in their enterprise, if only out of gratitude for what the administration had done for them during the dark days of 1943–4. He was disappointed.

They felt in the meanwhile that since Rommel had been chased out of North Africa they were secure in their concession and more disposed to thumb their nose at us.[14]

Secretary Ickes made one last attempt to get the US government involved in the Saudi Arabia oil bonanza. Calculating that if the front door had been slammed in his face there might still be another way in by a side entrance, he came up with the suggestion that the US should finance and build a pipeline from the Persian Gulf to the Mediterranean, thus saving the long tanker-haul by way of the Gulf, the Red Sea and the Suez Canal. In return for its outlay in building the pipeline (which was estimated in 1944 at $120,000,000) the government would receive all the oil it needed at 75 per cent of market price.*

This time the secretary's proposal was embraced with enthusiasm by the Arabian oil company's directors, who knew

---

* The scheme was not new, but had simply been appropriated by Ickes. It was first proposed by a US admiral (Andrew Carter) and later approved by the joint chiefs of staff, and James F. Byrnes, director of the office of war mobilization, recommended it to the president in January 1940.

a bargain when one was waved under their noses. Not only would they get a pipeline for their products without putting out a cent, they would also get one quickly, for the government would give priority for the project and raw materials (steel for the pipes, for instance) at a time when they were in desperately short supply. But once again the spectre of government involvement in the oil business was too appalling a vision for the rest of the industry to endure, and such formidable opposition to it was built up inside the United States (about 98 per cent of the industry was against it, according to W.S.S. Rodgers of the Texas Company) that, as Ickes himself said when writing about it later, the scheme was 'done to death without benefit of clergy'.[15]

Aramco decided to build a pipeline on its own (at vastly greater expense) and it went into operation on 28 January 1949, after agreements were signed with Saudi Arabia, Syria and the Lebanon, the countries the pipeline traversed on its way to the Mediterranean. When one considers the way in which Tapline (Transarabia Pipeline) has been used ever since as a blackmail weapon by the Syrian government and Palestine guerrillas,* it is fascinating to think of the mess in which Washington might have involved itself had the pipeline belonged to the government instead of to a private company.

* Its flow has been cut or part of the line blown up fourteen times in the last five years, usually as a protest against US government policy towards the Arabs and Israel.

12 · Breaking the Thick Red Line

Once the US government had agreed to respond to the oilmen's plea and bail them out in Saudi Arabia, they came to the rescue with golden buckets. Two military missions flew in during 1943, the first led by General Royce and the second by Lieutenant-Colonel Hoskins. Supplies followed in their wake: desert trucks, bulldozers, road-making materials, and the job begun of turning Saudi Arabia's rock-strewn desert tracks into some semblance of roads. Early in 1944 the first resident US minister was accredited to the kingdom in the person of Colonel William A. Eddy, a shrewd and monominded patriot, son of a missionary, born and brought up in South Lebanon, a fluent Arabic-speaker. At the same time Dhahran, which was rapidly quadrupling in size, thanks to the labours of thousands of Italian prisoners-of-war, had its first diplomat in the person of a US consul general, who opened his offices on a small jabal overlooking Dammam Number Seven.

As if to recompense the oilmen for having failed to provide them with a government-built pipeline, the president encouraged the Department of Commerce to heap other kinds of largesse upon the company. The benefits did not come in the form of grants of money but in something which, in those final days of war and early days of peace, was much more valuable – priorities for raw materials. All over the Western world everyone was now fighting to get hold of materials for rebuilding shattered factories and cities and for getting peacetime industry restarted. In the oil industry in the United States, and in British oilfields in Iraq, Iran and Kuwait, men were waiting for the machines and the tools that would enable them to restart old wells and drill new ones, to be ready for the postwar boom in petroleum that everyone knew was coming. But all of them had to wait on Saudi Arabia, which was now the US government's favourite son. By 1944 Aramco had its old fields going full blast again and was drilling for new ones. But more than that, and thanks to Washington's generous allocations, it had also started work on a vast new refinery at Ras Tanura, an undersea pipeline from Dhahran to the refinery on Bahrain Island, and—biggest undertaking of all – the Tapline project to the Mediterranean. For this latter undertaking, the Department of Commerce granted an export licence for no less than 20,000 tons of steel. When there was an outcry from independent oil producers, who complained bitterly that some-one in the Cabinet was unfairly favouring oil interests in Arabia, the Senate Small Business Committee called hearings on the subject, and heard from military spokesmen that they could not support the export grant. The Independent Petroleum Association produced a long and virulent attack on it, and was rewarded by a report from the Committee declining to recommend the export licence on the grounds that they could not find it in the public interest.[1] That delayed the matter for nine months, but at the end of that time the Department of Commerce quietly reactivated the export licence, and shortly afterwards the 20,000 tons of steel were on their way.

While Aramco was enjoying these priceless fruits of US government bounty, other oil companies in the Middle East could only look on and gnash their teeth in frustration. Though oil began to flow in ever-increasing quantities from Saudi Arabia from 1944 onwards, Kuwait, for instance, though only two hundred and fifty miles north of Dhahran, did not get started until 1947, and in Iraq and Iran expansion programmes took even longer.

The summit of the US government's love affair with Saudi Arabia was probably reached in 1945, when President Roosevelt and King Ibn Saud met for a Middle East conference aboard a US battleship. The object of the meeting, according to US spokesmen, was the president's hope of enlisting the king's influential support in the Arab world for a peaceful solution to the Palestine problem, which was then, as now, bedevilling the Middle East. That it was more an exercise in the consolidation of US–Saudi relations, to the undoubted benefit of American oil interests, is indicated, however, by the secrecy with which it was arranged.

It was not until the last day of the Yalta Conference in December, 1944 (when Roosevelt, Stalin and Churchill met in the Russian Crimea to parcel-up spheres of interest in liberated Europe) that the president told the British prime minister that he was going to see King Ibn Saud on his way home. According to Colonel Eddy, the US minister in Jiddah, who had fixed the presidential meeting with the king, Churchill was 'thoroughly nettled' and 'burned up the wires of all his diplomats' with orders to arrange a similar meeting, but was too late to get there first.[2]

The meeting between king and president took place aboard the US cruiser *Quincy*, which was anchored with President Roosevelt aboard, in the Great Bitter Lake, in the Suez Canal area, and Ibn Saud was transported there from Jiddah by the US destroyer *Murphy*. To get to Jiddah from his capital at Riyadh the king set off in mid-December, 1944, with a cavalcade of two hundred cars and almost at once, in the 600-mile trek across the desert, ran into violent winter rains and flash-floods which bogged down the caravanserai for a week. The king had grown old in recent years, and had taken to dyeing his beard and his hair, but as if to insist that his virility was unimpaired, he now refused to travel without his harem, and brought them, muddy but otherwise unharmed, into Jiddah at the end of the month. Eddy was horrified when he was informed by court officials that not only did the king expect to take two hundred of his retainers with him aboard the *Murphy*, but that he planned to have a selection of the beauties of his harem travelling with him. The US minister had a message in his pocket from the captain of the *Murphy* informing him that the *Murphy* could not possibly accommodate more than the king, four advisers, and eight servants. With visions of what would happen if a bunch of veiled Arabian houris were suddenly introduced into a complement of muscle-bound US sailors, Eddy went to work to acquaint the Saudis with the realities of the situation.

Ibn Saud finally compromised on a party of forty-eight – all males. They consisted of, among others, his second son, Prince

Faisal,* Abdullah Sulaiman and two other ministers, his private physician, his chamberlain, his coffee-servers, his cooks, and six enormous Nubian slaves armed with swords. A US destroyer has no accommodation for kings, but the Saudi party covered the quarter deck with an Arab tent, placed rich carpets on the deck, and erected a throne for their monarch. There was a critical moment when a courtier tasted the ship's distilled drinking water and pronounced it 'dead', and sent ashore for supplies of Mecca well-water (the only other well from which the king would drink being in Riyadh); but the worst crisis occurred just before the *Murphy* sailed out of Jiddah, when an Arab dhow came alongside with a cargo of eighty-six sheep, alive and braying. The destroyer's captain, who must have been wishing by this time that he could escape to something less nerve-racking, like fighting the Japanese fleet, tautly informed the Saudis that he could not possibly take live animals aboard his ship, and that there was plenty of meat for everybody in the *Murphy*'s amply-stocked refrigerators. *Dead meat*, responded the Saudis, *rotting carcases*. They insisted on their live sheep, to be slaughtered just before cooking. With the thought of having to pass guiltily before the accusing eyes of doomed animals every time he walked his own decks, and with his imagination picturing decks awash with the unfortunate victims' blood, the captain finally consented to have ten, but no more than ten, brought aboard.

The *Murphy* sailed, and luckily there were no more crises. The king saw a film of the naval war in the Pacific, while his sons sneaked away to join the sailors who were watching a rather more worldly comedy with a heroine in her underwear. There were practice firings on the way up the Red Sea, and five times a day, after solemn consultations with the navigator, the king and his party knelt and prayed in the direction of Mecca. As the destroyer neared its rendezvous with the *Quincy*, the king presented each crewman with $40 and each officer with $60, together with (for the officers) an Arab costume and a gold watch. The captain received a gold dagger, and in turn presented the king with a pair of binoculars and two submachine guns.

There were gifts given aboard the *Quincy* too, for King

* Now King Faisal, ruler of Saudi Arabia.

Ibn Saud, who had grown lame himself, admired President Roosevelt's wheelchair and was immediately presented with his 'spare'.* He also promised to send the king an airplane, big enough to carry himself and selected members of his harem, and therefore convenient for travelling between Jiddah, Mecca, Medina and Riyadh, now beginning to seem such long and weary journeys by car over bumpy desert tracks. In return the king gave the president jewelled swords, daggers and perfumes in richly decorated bottles, and while they talked of farming, and of US–Saudi co-operation in the extraction and marketing of Arabia's oil, they got on like old friends. On the solution of the Palestine problem they were rather less successfully harmonious, and their conversation built up to a misunderstanding that was to sully US–Arab relations in the months to come. For Roosevelt first verbally promised and then confirmed by letter that as president he would never do anything hostile to the Arabs, and that the United States government would make no change in its Palestine policy without consulting both Arabs and Jews beforehand. To King Ibn Saud this was a binding promise which would be kept as one made by himself; he did not understand the American form of government and believed that when a president spoke his word was law, and he no more dreamed that it could be altered by his successor, or by Congress, than his own decisions could be reversed by the people of Saudi Arabia. Two months later Roosevelt died, and his pledge died with him.†

In the circumstances, Winston Churchill's meeting with the king was rather less cordial than that of the president. It took place at Fayum, Egypt, three days later, and there was little that the British prime minister could offer Saudi Arabia and little that the king might give in return.‡ The Americans were

---

* It turned out to be too small for the giant-sized monarch, but he took it back to Jiddah with him and had a larger model copied from it.

† In breaking Roosevelt's promise, President Harry S. Truman used words which were, from then on, to haunt every oilman or diplomat trying to do business with the Arabs: 'I'm sorry, gentlemen, but I have to answer hundreds of thousands of people who are anxious for the success of Zionism. I do not have hundreds of thousands of Arabs among my constituents.'

‡ Except the ritual presents, of course. Jewelled sword and dagger, genuine Gulf pearls and other precious stones from the king; a set of perfumes from the prime minister and a promise (to match Roosevelt's airplane) of 'the finest car in the world,' a Rolls-Royce which Ibn Saud never used because it was a right-hand drive, and the

now paying his bills and were getting Arabia's oil in return. As for the fraught question of Palestine, Churchill believed it was a problem on which Arabs and Jews would never agree, and he could offer no solution and refused to make any promises about Britain's attitude. Moreover, if Ibn Saud was a proud and autocratic man, Winston Churchill was no less so, and as leader of a Britain newly emerged as victor in the war he was not going to allow himself to be outranked by a mere desert king. It was explained to him that Ibn Saud was a strict Muslim and neither smoked nor drank, nor did he allow either vice to be pursued in his presence. President Roosevelt, it was added, though a chain-smoker, had abstained in the king's presence except for a before-dinner cigarette which he smoked in the ship's elevator, to prevent the sinful odour from reaching the king's nostrils. Churchill was having none of that, and he pointed out to the interpreter that 'if it was the religion of His Majesty to deprive himself of smoking and alcohol, I must point out that my rule of life prescribed as an absolute sacred rite smoking cigars and also the drinking of alcohol before, after and if need be during all meals, and in the intervals between them.' The prime minister added that 'the King graciously accepted the position,' but the fact was that he was away from his own country and had no option.[8]

King Ibn Saud returned to his capital to be greeted by a great crowd of cheering Saudis, proud of what their ruler had brought back from his meeting with Roosevelt, materially and politically. The material gains were more durable than the political promises, for they brought technical aid to Saudi Arabia on an even more generous scale than before, and an assurance that from now on the oil from Arabia's wells would flow in ever-increasing volume.

In the first year after the end of the war, Aramco paid out $20,000,000 in petroleum royalties to King Ibn Saud. Soon the payments would grow to $4,000,000 a week. The money was handed over directly to the king, and it came to him at a moment when he was beginning to lose his grip. It was about this time that Harry St John Philby returned to Saudi Arabia

---

king, who liked to sit in the front of a car, refused to sit on the left of the driver, a demeaning position for a superior Arab.

from an enforced exile in wartime England and he found the king a sorry shadow of himself.

He was a careworn man [he wrote], already tiring under the strain of a vigorous life: with a crippled knee to which he was rapidly surrendering, and other signs of the inroads of the great enemy, which the most skilful dyes could not conceal. He was clearly following the path of least resistance through the dark forest of a world he had never seen and only knew by hearsay ... the quick clear mind was blurred and hesitant as it struggled with the problems of a strange world.[4]

He now had in his hands more money than even he had ever dreamed of, and there was a whole nation waiting to gain some benefit from it. But somehow it seemed to trickle through his fingers, into the pockets of the profligate hangers-on around his court. The wealth that should have been the making of Saudi Arabia was soon threatening to be the ruination of it.

Now that the United States government was no longer building the transarabian pipeline for Aramco, the company had to find the money to pay for it, and for the new refinery at Ras Tanura and the new undersea pipeline to Bahrain. This vast programme was far beyond the financial resources of the company, and at the beginning of 1944* Aramco began negotiations with two other US petroleum corporations, offering them a 30 per cent and a 10 per cent participation in Aramco respectively. These companies were Standard Oil of New Jersey and Socony Vacuum Oil Company, and they were what Standard of New Jersey's official history calls 'obvious prospects' to answer Aramco's needs. 'They could command large capital and were willing to assume the risks. Both needed more oil for their markets – Jersey especially, for its outlets in Europe.'[5] If the deal went through, Standard Oil of California, the original exploiter of the concession, and the Texas Corporation, which had joined California in 1936, would each retain 30 per cent participation in the new quadrumvirate, and Saudi's oil would be in the hands of what were now four of the biggest, and most competitively aggressive oil companies in the world.

But before the merger was consummated, there were to be

* Shortly after the company changed its name from California Standard Arabian Oil Company (Casoc) to Arabian American Oil Company (Aramco).

four years of hard arguing, legal threats, backroom meetings
and thousands of miles of travel by the chiefs of the companies
involved. For there was a snag – and it brings Calouste
Gulbenkian, otherwise Mr Five Percent, back into the story,
angrily waving a map of the Persian Gulf area with that
famous red line drawn around it. Because, of course, Standard
of New Jersey and Socony Vacuum were still signatories of the
Red Line Agreement, and it applied to Saudi Arabia which had
once been part of the old Turkish Empire. This meant that
neither company could join Aramco without permission from
their fellow members of the Iraq Petroleum Company, and
though that might be forthcoming from some of them, their
number did not include Gulbenkian. He would want his
inevitable five per cent of any new deal that the Red Line
signatories made. How to circumvent him?

Calouste Gulbenkian had prospered, thanks to the Red Line
Agreement, and he was now well on the way to becoming the
richest man in the world. From IPC and other enterprises in
which he was now involved he was receiving at least
\$20,000,000 a year. He and his wife had moved to Paris in the
1930s but there, except for keeping up appearances, went their
separate ways. Theirs had been an Oriental marriage arranged
by their families, and Nevarte Gulbenkian, plump and pretty,
adored by her brothers, had never really liked the ugly, humourless husband who had been forced upon her. Gulbenkian gave
her a house in the Avenue d'Iena and visited her there at least
once a day, usually for the midday meal. Once a week he carefully checked her household accounts and details of how she
had spent her allowance, and she often had difficulty disguising
the sums spent on presents for her succession of lovers; but
there was never any question of a divorce between them. The
supposition that they would remain married was such that
Gulbenkian housed his formidable art collection in the Avenue
d'Iena house, and often brought art collectors and curators from
all over the world to see his paintings and sculptures there.*

Gulbenkian preferred to live in his permanent suite in the
Ritz Hotel in the Place de la Concorde,† and it was there that

---

* They included some chocolate box nudes of the type oil millionaires seem to like,
but also, thanks to his art adviser, Sir Kenneth Clark, Cezannes, Renoirs, Monets and
several works by Degas.

† He was a director of the company owning the hotel.

he indulged a sexual appetite that was almost as voracious as that of King Ibn Saud, though he did not use up anything like as many women in satisfying it. As the years went by, however, his mistresses became younger and younger until, towards the end, nothing but a Lolita would stimulate him.

He used to say, and the late Lord Evans agreed in this [wrote his son, Nubar], that while it is very unkind on a young girl to have sexual relations with an old man because she loses her youth, it does rejuvenate the sexual functions of an old man. This was always reognized in the harems of the East and even today at the court of King Ibn Saud of Arabia or that, until his death, of the Pasha of Marakesch, the harem contains one or two young girls just past the age of puberty who are kept, and regularly replaced, for precisely that purpose.[6]

He was miserly with his mistresses until such time as he was sick of them, when he pensioned them off generously enough, to keep them quiet. He was so determined to keep hold of his possessions that his son tells the story of a rolling pin which Gulbenkian provided for his Turkish cook with which to roll out the wafer-thin Oriental pastry he was fond of. When the cook departed to go to work for the Turkish ambassador in Moscow, he took the rolling pin with him. It was worth only a few pence, but Gulbenkian wanted it back and sent two detectives to interview the chef in Moscow and demand its return. He got it.

Though his son, Nubar, was his most trusted assistant and had acted as his emissary and negotiator in some complicated deals (including the Red Line Agreement) he continued to treat him as a wayward boy, and would stop his allowance or cut it back if he were dissatisfied with his son's caprices. He was both mean and greedy, and ever watchful of his erstwhile partners in the oil business, in case they should try to do him down (as, admittedly, they often did). He had few friends, and none of them in the oil business.

'Oil friendships are greasy,' he used to say. He really believed only in one thing, and that was the power of money. Money came to him through the Red Line Agreement, and he clung to it as if it were his lifeline.

But then he fell into a trap.

When France was overrun by the German Army in 1940 Calouste Gulbenkian did not flee the country, but moved instead with Marshal Petain's government to Vichy. The Vichy administration, after signing an armistice with Germany, subsequently broke off relations with Britain and came under Nazi domination. At once the British custodian of enemy property declared that he was taking over Gulbenkian's interest in IPC and that of the Compagnie Française des Petroles, since both of them were now subject to enemy control. Their shares were declared forfeit. There was little that the French could do about it, but Gulbenkian was furious when he heard the news, for he held an Iranian as well as a British passport, and was free to move had he so wished. In fact in 1942 he moved to Lisbon, in neutral Portugal, and fought strenuously and successfully to regain his 'rights' from the custodian and be paid compensation for his loss of revenue.

But when, soon after the end of the war, Standard of New Jersey began to search for a way of breaking the provisions of the Red Line Agreement, which was holding up its participation in Aramco, the wartime confiscation suddenly pointed the way. Jersey consulted their legal expert, who advised that, in his opinion, 'the fact that two of the owners of the Iraq Petroleum Company – Compagnie Française des Petroles and an individual, Calouste Sarkis Gulbenkian – had come under enemy domination by the German occupation of France in 1940 had terminated the working agreement of 1928.' As a result, the counsel held, the Red Line restrictions were no longer in force.[7] To obtain a broader judgement, the head of Jersey's law department, Edward F. Johnson, also consulted several eminent counsel in London, including a famous lawyer named D. N. Pritt.

He wanted their advice concerning the position that might be taken on this issue by the British courts, which under the bylaws of Iraq Petroleum had jurisdiction over matters in dispute between its owners. These British lawyers were unanimous in the opinion that the Red Line Agreement was no longer binding on the signatory companies.[8]

Jersey was also looking for other allies to bolster its case, and found one in the US State Department. From officials there it secured an assurance 'that the Government would not support a new agreement incorporating the old restrictive features' and was against 'the hobbling of American nationals by such devices as the Red Line Agreement,' and that the Department stood for 'what was essentially an opendoor policy in the Middle East, but with safeguards for the interests of the producing countries.'⁹

Armed now with some heavy ammunition, Orville Harden, a vice-president of Standard of New Jersey, and Mr Will Sheets, an official of Socony Vacuum, arrived in London with instructions to sink all opposition and wipe out the Red Line Agreement once and for all. They had only slight trouble with the two principal British signatories to the agreement, Royal Dutch-Shell and Anglo-Iranian, for they had taken counsel's opinion also and agreed that the war had terminated the Red Line Agreement. But Anglo-Iranian knew that once Standard and Socony Vacuum were part and parcel of Aramco, that company would open all stops to get oil flowing, and this American flood might well affect Anglo-Iranian's own growing sale of crude on the world markets.

Orville Harden had already been primed over how to deal with that one, and what followed was a good example of the master strategy of which Standard of New Jersey was capable when its board got down to serious planning. First of all Harden negotiated a deal with Anglo-Iranian whereby his company would purchase from the British a large amount of crude (it was eventually fixed at 110,000 barrels a day) for twenty years beginning in 1952. He also agreed to participate in a new corporation to build a new pipeline from Iran and Kuwait to the Mediterranean if the building of such a line were found feasible. These arrangements more than appeased Anglo-Iranian's anxieties about future competition in the crude oil markets, and ended any doubts they might have had about the elimination of the Red Line.

From Standard of New Jersey's point of view, the arrangement was a double achievement. It not only took care of British objections to its plans; it also enabled Standard to proceed with a cherished scheme of its own to establish itself solidly in the British petroleum market for private cars. In this period of postwar austerity, Britain's petrol supply was still under government control and private motorists were stringently rationed. Whenever public opinion and the newspapers demanded an end to the rationing system, the Labour government cited as their reasons for refusal the fact that the bulk of the petrol would have to come from American sources, and would have to be paid for in American dollars, of which Britain was desperately short.

These statements rankled with Howard W. Page, the resident Jersey director in London. 'It made the American companies look like the bad boys,' he said, 'and it was hurting us because the public thought we were the dirty dogs who were keeping their family cars off the roads.'[10]

So while nursing a broken ankle from a skiing accident, Page worked out a scheme.

'It took me two months to work it out with New York,' said Page, 'but, thanks to government help, I got it through the British Parliament in two weeks. It was a very simple scheme. With any increase in consumption of our oil, we announced, we would take payment in 100 per cent sterling. And we would use that sterling for the purchase of British goods for use around the world. After the government's statement that ours was dollar oil and theirs was sterling oil, and they could only pay for sterling oil, this made ours sterling oil in the definition of the term. The Labour government was furious at first. They sent guys from the Treasury to try to explain that it wasn't as simple as that, but I told them I didn't understand what they were talking about, and they finally gave up. So did the government. They knew and I knew that 99.8 per cent of the people were for the scheme and only .2 per cent of politicians and black marketeers were against it, and it had to go through. The scheme was officially accepted, and petrol rationing ended in Britain. Of course both the Labour government and the Tories took the credit for it, but we didn't mind that.'[11]

By making its agreement with Anglo-Iranian, Jersey now didn't even need to supply Britain with dollar oil because it had arranged for ample supplies from a sterling source. So Anglo-Iranian's agreement to waive the Red Line restrictions had brought them a double benefit.

Since Royal Dutch-Shell had also waived its rights under the agreement (in return for similar crude-oil contracts from Socony Vacuum), there were now only the French and Calouste Sarkis Gulbenkian to deal with. Gulbenkian was now permanently ensconced in a hotel in Lisbon, but his son, Nubar, was authorized to act on his behalf. He was in Paris conferring with Victor de Metz, president of the Compagnie Française des Petroles (the other main shareholder in IPC and signatory of the Red Line Agreement) when the two men were summoned to London for an urgent meeting with Harden and Sheets. There they were told that the Department of Justice considered the Red Line Agreement of 1928 a restrictive agreement, contrary to American anti-trust legislation, and that they therefore no longer considered themselves bound by it. Nubar at once pointed out that what the US Justice Department thought about the agreement was hardly relevant, since all the signatories had agreed that any dispute over it should be settled under British law. The Americans then changed their tack. From their briefcases they produced the opinions of various eminent British counsel that, owing to the occupation of France, both Gulbenkian and the Compagnie Française des Petroles had come under enemy domination, and agreements with them were no longer binding.

De Metz looked at Nubar Gulbenkian, who seemed quite unfazed by this thunderbolt.

'On va voir,' said de Metz, softly, and the meeting was adjourned. Some days later, through their legal advisers, the French and Gulbenkian formally demanded a share under the Red Line Agreement in Jersey's and Socony Vacuum's projected interest in Aramco. When this was refused they announced (in February 1947) that they were starting legal proceedings 'to obtain confirmation of the validity of the 1928 agreement and a declaration prohibiting [Jersey and Socony Vacuum] from obtaining an interest within the Red Line Agreement independent of the other owners of Iraq Petroleum.'[12]

The news created something like panic not only in Jersey's board-room in New York, but also at Aramco, where all their plans were now threatened. In March 1947 Jersey and Socony were forced by the impending action to work out 'standstill' agreements with Aramco and its owners.

The purchase by the two American companies of participation in Aramco and Tapline was to be held in abeyance pending clarification of the legal issues.[13]

Aramco had the worst headache of the lot. In anticipation of the sale of stock to the two newcomers, it had gone ahead with its plans, and it just did not have the money for them. However, after several emergency meetings,

The two prospective buyers agreed to guarantee a bank loan of $102,000,000, this being the total amount they were to pay directly for their shares in the company. They also worked out with Aramco and its two corporate owners the other terms of the purchase, which included their foregoing of specified amounts of dividends for a period of years. At the same time they arranged to purchase oil from Aramco in the interim and to guarantee their proportionate share of a loan for Tapline of $125,000,000.[14]

Undoubtedly, the first round had gone to the French and the Gulbenkians.

From this point onwards, there were too many backroom meetings, bribes, threats and blandishments to be able to go into them in any detail here. At all costs the Americans and the British signatories wished to avoid a legal action over the Red Line Agreement, for they knew that the Gulbenkians' lawyer, Sir Cyril (now Lord) Radcliffe, was a brilliant performer in this type of case, and they feared the dirty domestic linen that might be aired, thanks to the material which Nubar and his father could give him.*

Therefore, the tactics they first of all adopted were to wean the French away from the Gulbenkians. These were the lean postwar years for the French, and they needed to sell as much oil as possible as a boost to their economy. Their only oil at that time came from the wells of the Iraq Petroleum Company. Suddenly the French heard reports that IPC proposed to cut down on its production, which would mean that Aramco and Anglo-Iranian would benefit at the expense of IPC, thus drastically reducing France's income at a moment when she was desperately in need. There was nothing that the French could do about it either, since their British and American partners outvoted them. At the same time as this threat was waved over their heads, however, the French were offered a carrot. Since they had lost their shares in IPC to the British Custodian of Enemy Property during World War II, they had naturally lost the revenues which had been earned from them at the same time. It will be remembered that the astute Gulbenkian, claiming that he had been deprived under false

---

* In a case which Nubar once brought, Radcliffe compelled the defendant (who happened to be Gulbenkian père on that occasion) to produce 987,000 documents weighing a ton. Nubar was suing for an increased allowance from his parsimonious father. He got it.

pretences, had got his back. But there was a genuine legal doubt in his case. With the French, there was no doubt whatsoever that the seizure had been perfectly legal. On the other hand, the British and American shareholders indicated their willingness to advise the British government to make a gesture to the French – providing, that is, that the French were willing to be cooperative in return. ...

It was at this stage that the French government sent to London a new representative from the Compagnie Française des Petroles and told him to make a compromise with the British and Americans, with Gulbenkian's agreement, if possible, but without it if the Armenian remained adamant. It so happened, however, that the new French emissary was Robert Cayrol, who had enjoyed the friendship of Calouste Gulbenkian for many years, and got on extremely well with his son, Nubar. It was Nubar who persuaded him not to hurry things. 'The time when others are pressing you for an urgent decision is the time to take it slowly,' he said, quoting his father. It was the Americans who desperately needed the agreement, he pointed out, and every delay was costing them money, ideal circumstances in which to drive a hard bargain with them.

We were both busy men so that most of these discussions [with Cayrol] took place at the end of the business day [wrote Nubar later]: Cayrol used to come round at about seven o'clock—to the Ritz when I was in London or to the Georges Cinq when I was in Paris—and we would have a good dinner together. There was no question of ice-cold negotiations but very friendly ones in which we sought a solution acceptable to all parties: true, we were careful to drink the same amount of wine to ensure parity. Session by session we worked gradually towards a basis for agreement until at last we reached the stage at which we felt it worth while to bring in the others.[15]

The Gulbenkians had realized by this time that there would have to be some relaxation in the restrictive powers of the Red Line Agreement, but both they and the French were determined to make it plain to the Americans that release from its provisions could only be obtained at a stiff price. So long as that price was not forthcoming, the threat of legal action would persist; and, in fact, the Gulbenkians instructed Sir Cyril Radcliffe to proceed with preparations for the case as if it were going through. A date was actually set in the Law Courts in London for the hearing. In the meantime, Nubar and Cayrol worked on a formula that would produce for them both *more* and not less revenue when the Red Line was finally erased.

At last, in November 1948, negotiations with the British and Americans had reached a stage sufficiently advanced for all parties to assemble in Lisbon, where, the Americans fervently hoped, an agreement would finally be signed and they could go ahead and join Aramco. The meeting place was the Aviz Hotel, where Calouste Gulbenkian had installed himself ever since leaving Vichy France in 1942. It was a rococo palace with chandeliers hanging from high ceilings, statues of maidens crouching between marble pillars, walls tapestried with hunting scenes, guests with kingly or princely titles (usually in exile), and servants who spoke only in whispers. Calouste Gulbenkian lived in suite number 42, and though, in effect, the mountain had come to Mohammed through this meeting, Mohammed did not deign to put in an appearance during the course of it. He sent the oilmen messages through his emissary, Nubar. Cayrol was there representing Compagnie Française des Petroles, Orville Harden and Howard Page for Standard of New Jersey, Sheets for Socony Vacuum, and Morris Bridgeman for Anglo-Iranian. There was also a representative of Royal Dutch-Shell. All of them were anxious to have the signing of a new agreement go through as urgently as possible, because the Gulbenkians' action was not only still in the British legal calendar but was due to come up in court at any moment, 'and it was touch and go whether the case would open in the Law Courts in London before agreement was reached formally in Lisbon'.[16]

The agreements were finally being typed out and were to be ready for the formal signing at seven o'clock in the evening. It was Sunday 14 November, and the Gulbenkians' English lawsuit was due to begin the next morning, but arrangements had been made to telegraph London and announce a settlement as soon as the signatures were on the papers.

It was at five minutes to seven [wrote Nubar], that father found one more point which had not been covered by the Agreements. To say that there was consternation on the faces of all the men gathered there would be a piece of English understatement. But father was determined. Telegrams were sent to London, where the unfortunate

Boards of the Groups involved had been kept waiting on tenterhooks for news that the Agreements had been signed. Now they must consider a new point and, in turn, send telegrams to Lisbon giving their acceptance or otherwise of the latest Gulbenkian demand.[17]

Nubar had ordered a dinner to celebrate the signing of the agreements, and he had made sure that it was the best that the Aviz could provide. A gay, dandyish, bearded leprechaun of a man, full of high spirits and a sense of fun, he was never one to take business setbacks too seriously, and he suggested to the others that they might as well eat while they waited for the telegrams from London.

There were some twelve of us at that table [he wrote], all men, with the sole exception of Cayrol's wife . . . No doubt she had looked forward as I had to a gay, convivial evening, but if ever there was a gloomy occasion, that was it. The meal was accompanied by long periods of silence, for no one was inclined to make the conventional efforts at conversation when all our minds were turned to what might then be going on in London. I realized hardly anyone was drinking, either, and my impression was confirmed the next day when I settled the bill for the dinner: just one bottle of champagne had been enough for twelve people.[18]

It was not until the early hours of the morning that word came through at last that the British and the Americans had conceded Gulbenkian's last point. It was not until two in the morning that the agreements, newly retyped, were signed at last by everybody, including Calouste Gulbenkian.

Once more Nubar Gulbenkian called for the champagne, but it had been taken away and the Aviz kitchen staff had gone to bed. So the negotiators celebrated instead on sandwiches and cheap wine from an all-night cafe.

Nonetheless, it was a cause for celebration on everyone's part, even though it had cost the major shareholders in IPC a great deal of money. Henceforward, all of them could pursue their development plans free from the tethering effect of the Red Line, which now vanished from oil companies' maps. Anglo-Iranian and Royal Dutch-Shell had found new and lucrative markets for their crude. Gulbenkian and France had secured written assurances that the Iraq field under no circumstances would be restricted, but, on the other hand, would have its output considerably expanded; and Gulbenkian was also given an extra allocation of free oil, for selling on the open market, as well as his five per cent of the profits, from Iraq and from the new fields in Qatar. It was to add at least $8,000,000 to his annual income.

As for Standard of New Jersey and Socony Vacuum, they were free at last to take up their holdings in Aramco. By 1950 they and their two partners were masters of the richest oil quadrumvirate in the world. Oil was at long last beginning to flow from the fruitful new field at Abqaiq by pipeline across the Arabian Desert to the Mediterranean, 1,400 miles away. A new refinery was in operation at Ras Tanura, and there were berths there for a great tanker fleet, which now began regular shipments to Africa and the Far East. By 1950, two years after the Red Line was erased, Standard of New Jersey's 30 per cent share of Aramco Oil was 164,000 barrels daily, and its share of the estimated reserves no less than 2,800,000,000 barrels.

The future looked bright indeed, and everyone should have been happy. But two people were not.

The first of them was Nubar Gulbenkian. For three years he had used all the wit, charm and intelligence of which he was capable (and he had these qualities in plenty) to win his father's war with the giant oil combines. True, when the crunch came, it was his father's nerve and tenacity which counted, but the son deserved some credit for his untiring work on Gulbenkian's behalf. Not only did he get no more than a perfunctory thanks from the old man, but when he left to return to London he asked for his bill and realized that he had handed out so much hospitality to the other delegates that the amount came to far more than his father allowed him as expenses for the trip. He rushed to Gulbenkian's suite to complain, and finally persuaded the parsimonious old man that next time he came to see him in Lisbon he would be allowed double his current expenses.

Three weeks later, he came to see Gulbenkian again. When he called for his bill at the end of this visit, he discovered that the price of his room had been trebled. And it was not until after his father's death that he discovered that Gulbenkian owned the hotel.

King Ibn Saud was also unhappy. He had watched from afar the negotiations going on in London, Paris, New York and Lisbon, but no one had thought of consulting him in Riyadh. The future of his own country was being settled by foreigners and infidels, and he had played no part in it. He was old and in pain and beginning to lose his grip, but not enough to blind him to the fact that he had been ignored – and that was insulting.

He decided to react in the only way an infidel American would understand. He asked for more money. And he invited another infidel to help exploit Saudi Arabia's oil, in competition with Aramco. At which point an independent oilman named J. Paul Getty enters the Middle East story.

PART FOUR

Enter the Independents

13 · Neutral Zone

A few days after the big international oil corporations signed away the Red Line Agreement in November 1948, a small private airplane took off from a sand-strip just outside the town of Kuwait and headed south towards the desert. It flew across the rolling sand-sea for about fifteen minutes until it reached a range of small jabals that ran down towards the Persian Gulf, and then it came down low. For the next two and a half hours, while the bedouin tribes shook their fists and finally fired rifle shots at this noisy machine that was scaring their camels and their women, it cruised round and round in circles. Then, at a thumbs down signal from the pilot to the man with the instruments beside him, the plane turned and flew back with a near-empty tank to Kuwait.

The name of the pilot has disappeared from the records, but his passenger was a clever geologist named Doctor Paul Walton. That night he went to the Kuwait post office and cabled his boss: STRUCTURES PROMISING. And it was on the strength of that message that Jean Paul Getty came into the Middle East oil business.

J. Paul Getty was fifty-six years old in 1948, and he had reached a turning-point in his life. A tall, gaunt Middle-Westerner from Minneapolis (though his ancestors came from Ulster and Scotland), he had the long dewlapped face of a sad-looking lion. His melancholy expression was due more to his private life than to any doubts about the oil business. His fifth marriage (to the former Louise Dudley Lynch) was going through the separations that would eventually lead to a divorce, but he had reached the glum conclusion some time earlier that he would never have a stable marriage,* and when

* 'I blame my business interests for having been married five times,' he declared later. 'A woman resents a man being dedicated to his business. In fact, she resents anyone dedicated to anything but herself.'

this one ended he was already resolved to make different arrangements for female companionship. So far as the oil business was concerned, he had every reason to be pleased even if he was not yet satisfied. He was not yet a billionaire, but he had already shown the toughness and skill needed by an independent to play in the high-level poker game that was the US oil business. The big major companies and the successful independents have always fought with every weapon at their command (and they had plenty: secret boycotts, squeezes, market manipulations) to keep newcomers out of the competition, but Getty had beaten them all at their own game, and without giving away any of his overall control had become master of two of the biggest independent operations in America, Getty Oil Company and Tidewater Oil Company, even then worth nearly $400,000,000 between them.

There were those who thought of Getty as another Calouste Sarkis Gulbenkian, but so far as their respective roles in the oil business were concerned there was no comparison. True it was that both had fathers who were oilmen by adoption, who had launched their sons into the business world with generous gifts.* True it was that they both collected expensive works of art and had a taste for pictures and sculptures of plump and nubile nudes. They shared an intense interest in women and sexual athletics ('a man's driving force is sex,' Getty once said), a lifelong search for bargains of every kind, and a miserliness that made Gulbenkian check his wife's household bills and how his mistress spent her allowance, and Getty put payboxes on his guest telephones and rode around in four-year-old cars.

But Gulbenkian was a middle-man, a fixer, the graduate of an Oriental bazaar who was not really interested in oil, but only in the money it would bring him. J. Paul Getty was (and is) an oil man who likes the smell and feel of it. He has not only never grown rich on five per cent of other people's earnings, but has always been in control of every enterprise in which he has been involved. He prefers to compare himself with John D. Rockefeller, a visionary as well as an entrepreneur, who saw new ways of using oil before other men had thought of it, an innovator. Like Rockefeller, he kept a close check on every

---

* George Franklin Getty, a lawyer, took strips of Oklahoma Territory in lieu of fees and found himself landlord of an oilfield; he launched his son with a gift of $1,000,000 when he was twenty-one.

enterprise in which he was (and is) concerned, and explained his meanness by citing the stern mother who nagged him all through his formative years to remember that 'he who wastes not wants not!'

The end of World War II had found this humourless, lugubrious, lonely man walking around with what an acquaintance described as 'a look in his eye as if something is missing from his life'. It was not a woman, for he could always find plenty of willing companions, even though he would never waste a cent on them.* What he was missing was a stake in Middle East oil through which he could continue the kind of encounter which he enjoyed most of all in life: meeting a big business corporation head on and winning the subsequent battle. He had looked in Iran, Iraq and Egypt. In the first two countries the major oil companies had already got there before him, and in Egypt he did not think much of the prospects.

And then one of his paid informers quietly let him know that Aramco, once it had settled its future set-up with Standard of New Jersey and Socony Vacuum, was planning a major reshuffle of its Saudi Arabian concessions. As well as the areas which the company had leased at Dhahran, Abqaiq and Abu Hadrya, Aramco also held the concession in the so-called Neutral Zone between Saudi Arabia and Kuwait, this area having been part of the sale made to them by Major Frank Holmes back in the 1920s. Now they did a deal with King Ibn Saud whereby they relinquished all rights in the Neutral Zone in return for an offshore concession running out to sea from their Dhahran field. From Aramco's point of view, it was much more convenient, economically and administratively, to keep their concessions centred around one refinery and tanker port, and the Neutral Zone had complications.

The main complication was the fact that this two thousand square miles of rocky and inhospitable desert belongs not to one country, but two. When the British proconsul, Sir Percy Cox, had visited Ibn Saud at Oqair in 1922, one of the agreements he had secured between Arabia and the neighbouring territory of Kuwait was the establishment of a Neutral Zone along their frontiers where tribes from both would retain their grazing and watering rights, and no forts would be built. This meant that

---

* 'Surprisingly he is often like a little boy and brings out the mother instinct,' said one of them (Penelope Kitson).

both King Ibn Saud and Sheikh Ahmad of Kuwait had a say in what happened inside the zone, and Aramco figured that there could be contentions.

With the concession in the Neutral Zone relinquished, each ruler had the right to redispose of half of it for the highest price he could get. The Ruler of Kuwait got into the act at once by announcing that his fifty per cent would be auctioned off to the highest bidder, and that no 'derisory' offers of the type which had won prewar concessions would be considered. An American syndicate called Aminoil came forward at once with an offer which was so generous that the major companies decided not to compete, and Aminoil's bid was accepted. The syndicate was headed by Ralph K. Davies, who had once been a director of Standard Oil of California (as well as deputy petroleum administrator in Washington during World War II), and he was not only convinced that high prices would in future have to be paid for concessions but succeeded in persuading the fellow members of his syndicate to go along with him.

Nevertheless, they were shocked at how high the price turned out to be. For a fifty per cent share in the Neutral Zone's potential oil (and there was no positive proof yet that it was there in viable quantities) they paid $7,250,000 as an immediate down payment to Sheikh Ahmad and a guarantee of at least $625,000 in royalty payments each year. This was a far different affair from Kuwait concession proper, for the whole of which, in 1934, Kuwait Oil Company had paid out $170,000 plus a guaranteed royalty payment of $350,000. It was also different from Al Hasa in Saudi Arabia proper, for which Aramco had paid $200,000. Moreover, Sheikh Ahmad exacted several conditions which looked onerous enough to Aminoil and outrageously exorbitant to their major oil company rivals, who immediately and indignantly spread the word around that the syndicate was ruining the market. In addition to a royalty payment which worked out at twice the Kuwait rate, Aminoil agreed to pay 12½ per cent royalty on all natural gas sales, 7.5 US cents tax per ton of crude oil in lieu of taxes, promised to construct a refinery and give the sheikh a 15 per cent interest in it, to run an educational pro-gramme for Kuwait employees of the company, and build a new hospital in Kuwait town.

Moreover, since Aminoil's purchase was only fifty per cent

of the concession, no work could be started until the Saudi Arabian half was sold. Who would be prepared to pay the price that King Ibn Saud would ask, which was bound to be at least as much as Aminoil had paid to the Ruler of Kuwait? While the major oil companies were swallowing hard, and the main contender, Royal Dutch-Shell, was working out whether it could afford it, Paul Getty sent his lawyer, Barnabas Hadfield, to Jiddah with instructions to buy the concession from the king no matter how much it cost. It cost plenty.

First of all there was a down payment of $9,500,000. Secondly, there was an advance on royalties of $1,000,000, payable yearly, and not repayable if royalties failed to reach that sum. There was a royalty payment of 55 US cents per barrel (Aminoil was paying 35 cents), a pledge to deliver crude or paraffin or aviation spirit up to 100,000 gallons free of charge to the government, an agreed programme for schools and education, and an appointment of the king's own delegate to the company's board.

At their headquarters in London, New York and San Francisco, the directors of the world's major oil companies read the terms with a growing sense of dismay.

'This,' said Howard Page, of Standard Oil of New Jersey, 'could change everything in the Middle East.'

Indeed it could. On 20 February 1949, King Ibn Saud signed an agreement with J. Paul Getty's Pacific Western Oil Corporation giving him his half-share in the Neutral Zone concession, and was handed a cheque in return by Barnabas Hadfield for $10,500,000. It was a sum sufficiently high to make even a king thoughtful. If that amount of money could be handed over by a one-man company even before that company had brought up a barrel of oil, why had Aramco paid him only $28,000,000 dollars in 1948, a year when oil was flowing out of Al Hasa as never before and two of the richest companies in the world had now joined Aramco as partners? Why was Aramco paying him only 21 cents a barrel royalty when Getty was willing to pay 55 cents?

Since the Saudi Arabian exchequer was, as usual, empty and debts were mounting rapidly, the king instructed his minister of finance, Abdullah Sulaiman, to summon F. A. Davies, the resident Aramco director, to an emergency meeting. He wanted an answer to his question, and more money – much, much more money.

A year later he had it: $50,000,000 more. But thanks to the ingenuity of an American corporation lawyer, it was not Aramco which paid the bill but the US Treasury.

Even a semi-billionaire cannot just reach into his safe and pull out ten and a half million dollars, just like that. J. Paul Getty's assets were tied up in his oil and business interests in the United States, and he had to go to the banks for a loan to pay for his Neutral Zone concession. He therefore needed the money back as urgently as possible, and that meant getting wells drilled and oil flowing.

But it wasn't as easy as that. There are worse spots in the world than the Arabian desert in mid-summer, but not many. It becomes so hot that a man who stays in the sun too long can boil to death, his skin and blood vessels dehydrating under the intense heat. In the Neutral Zone there are only about four or five wells, and no underground reservoirs which can be tapped for water as there are at Al Hasa. Only flies and locusts, vipers and scorpions, seem to survive the intense heat of mid-desert, and along the coast the humidity is such that air conditioners spurt water and men bathe in sweat. When summer sandstorms come – and sandstorms are a feature of Kuwait – life is well nigh insupportable. And even winter, with its flash floods, its spectacular hailstorms, its sometimes bitter cold, can make existence a burden and a trial on the nerves.

In these circumstances, the set-up in which J. Paul Getty found himself in 1949, and for the next few years, was not exactly propitious for the speedy return of his money. For a man of his parsimonious nature, the clause in the concession contract which nagged him most was the one which obliged him to pay a million dollars a year to King Ibn Saud, oil or no oil. He had of course figured on delays (and large financial outlays) while water was laid on, communication lines constructed, rigs, drills, crews and air-conditioned living quarters shipped in. What he had not included in his calculations were incompetence, inexperience and stubbornness.

He discovered that his mandatory partners in the Neutral Zone concession were afflicted with all these vices. Aminoil, which held the other 50 per cent of the concession, was a consortium of ten different groups, the largest of which, Phillips Oil, held only 33 per cent of the company's shares. Aminoil's president, Ralph K. Davies, owned only 8 per cent of the action, and had been nominated by the others as a front-man because he was a lawyer, a former Standard of California vice-president, and a wartime controller of US petroleum supplies. They were all qualities for successful negotiating in board-rooms and government offices, and Ralph Davies was a charm-ing and effective operator on his home ground, somewhere within walking distance of the Fairmont Hotel or the White House. But he was less successful coping with the deviousness of Arabian oligarchs, or making vital and costly decisions in the gritty, boiling blackness of a Neutral Zone sandstorm.

Though Getty was complete master of his own half-share in the concession,* and could therefore speak in a stronger voice than any member of his partners' consortium, he allowed Aminoil (through Ralph Davies) to take control of the prelim-inary stages. It was a costly mistake. So was his decision not to go to the Neutral Zone himself but to send as his representative his eldest son, George Franklin Getty II. George Getty was twenty-five years old and had had a year at college after leav-ing the army. He had neither the years nor the experience to make his presence felt, either with the Arabs or with Davies, who was apt to wave him away like a persistent desert fly. To backstop his son, Getty had dispatched to Kuwait his two geologists, Paul Walton and Emil Kluth. From surveys they had made they were convinced that a commercial oilfield would be found in the west of the zone, where there was evidence of Eocene limestone formations. They recommended that opera-tions begin there. But Davies listened to his own geologists, who were after deep-hidden oil along the same seams as the rich Burgan field, just over the line in Kuwait proper. Unfortun-ately, their first wells were east of the Burgan line and came up dry, or full of salt water. And since they were deep wells, each failure cost a quarter of a million dollars.

In an effort to persuade Davies that the whole strategy of the campaign was wrong, Getty flew in an expert geophysicist from Anglo-Iranian, but his advice was ignored. Had Getty flown in himself, his driving personality would almost certainly have prevailed and they would have 'drilled where I wanted'

* He owned 82 per cent of the shares in Pacific Western.

and 'found the field three years earlier'.[1] But he was involved in his tangled domestic affairs and considerable stock market manipulations, and he had to stay close to a telephone and teletype. He could neither telephone the Neutral Zone nor send cables except *en clair*, and four years passed while Aminoil poured away their money and Getty's without any tangible result. By 1953 Getty had spent $40,000,000 in the Neutral Zone without getting a cent in return, and there were strong rumours in the oil world that he was ready to pull out. Towards the end of that year, however, a viable field was discovered inside the Neutral Zone but about twenty miles from the Burgan Zone in Kuwait proper. It was where Aminoil's experts had always said it would be, and to that extent they were justified; but the wells that were subsequently sunk were deep, enormously costly to maintain, and nothing like the bonanza for which Getty had been hoping.

It was not until an unlucky incident brought George Getty home (he had run foul of the Saudi liquor laws) that J. Paul Getty was forced to visit the zone himself. He drove out from London by car, which he co-piloted with his close friend of the period, an Englishwoman named Penelope Kitson, and immediately on arrival he galvanized everyone. He was scathing about the laxness of Aminoil's operations and administration. He stiffened the backbones of his own demoralized staff, and both impressed and terrified everyone who met him. 'Even the jellyfish seemed to get out of his way when he took a swim in the Gulf,' one oilman said. 'He was the only man I ever knew who bathed every day and never got stung once. The jellyfish wouldn't have dared – Getty would have stung them right back!'

As a result of this visit, the drilling tactics changed and rigs went out to new locations. And within a year, the bonanza for which Getty had been hoping bubbled to the surface of the Neutral Zone. At depths of only 600–1,500 feet, in seams of Eocene limestone, oil was found in abundance. It was soon being pumped out at the rate of 16,000,000 barrels a year, and the drills were biting into a vast oil reserve of approximately 13 billion barrels.[2]

It had cost J. Paul Getty $40,000,000 to tap it but it was to make him a billionaire.

14 · Fifty-fifty

From poles and baskets in the bazaars of Dammam, Riyadh, Hufuf and Jiddah the severed hands and feet of petty thieves still hung, flyblown and festering, in the blazing sunshine. They were not the only things that were rotten in the Saudi Arabian state. It was 1950 for the rest of the world, but the desert kingdom was a mediaeval despotism where sheikhly intolerance and religious fanaticism still held sway. Aramco wives were startled when they joined a crowd in Dammam watching two black slaves beating a sack, and heard anguished female screams issuing from it; and their surprise turned to horror when blood began to leak from the sack and they realized that they were watching a wretched female being ritually beaten to death. She was inside the sack so that the male watchers should not see her face. There were still strokes of the lash for persistent non-attenders at the mosque, and a man lost his head to the executioner's sword for stealing a camel or a man's wife.

In the early days of Ibn Saud's reign the system had been savagely cruel, but at least it had been egalitarian, and alongside the savage punishment for wrongdoers there was also an honour system based on ancient desert chivalry. But that was now beginning to decay and there was a smell of putrefaction in the air. King Ibn Saud was now going downhill mentally and physically, and he had lost his grip on the princes and ministers who grazed on the rich pastures of his patrimony. They sensed his imminent demise, and were beginning to make 'other arrangements'. In these last months of his life, a Dutchman named Dirk van der Meulen, an Arabist and an admirer of the mighty warrior-monarch in his heyday, came to visit him in Riyadh, and was received in audience at the palace.

When I entered the hall I saw a change [he wrote afterwards]. The

King at the far end no longer rose from his seat in welcome. Only the guards in Western uniform saluted. The bedouin Shaikhs near the entrance moved silently backwards to make room but gave no greeting. No one rose from his seat because the King could no longer do so. He sat in what seemed to be an invalid carriage. The spare, curled beard and the few locks of hair that peeped from his head-cloth were black—but dyed . . . It was the voice that disappointed most. The voice was still kindly but the music had gone out of it . . . When I left the audience chamber of Ibn Saud I felt, for the first time, unsatisfied. The unfailing spring had failed.[1]

What had set him on the road to indifference and death, it was said around the courts, was the failure of that strength which he admired in himself on a level with his skill and bravery in battle: his prowess as a sexual contender. But suddenly the wick had burned down. He had been stimulated for a time by a visit to Egypt, where his eyes glinted over the women of Alexandria, and he said to his Anglo-Irish friend, Harry St John Philby: 'There are some nice girls in this country. I wouldn't mind picking a bunch of them to take to Arabia, say a hundred thousand pounds' worth.'[2]

He did in fact bring three or four back with him, and, en-livened by new faces and new bodies, he had enjoyed an Indian summer of sex in which he fathered two more sons to add to his formidable brood of children. But they were the last. Suddenly he could father no more, and it produced a decline, it was said in Riyadh, that nothing could halt from now on.

In his feeble state he did nothing at all to check the corruption, the waste and the wanton profligacy with which his country's enormous revenues were now dispensed by his sons and his ministers. All the ministers now had Lebanese and Syrian secretaries whose only job was to siphon money out of their budget allocations and invest it for them in Beirut real estate or Swiss banks. So much Joy perfume was being imported into the country ('the most expensive perfume in the world') that one of the company's salesmen said: 'They must be taking baths in it.' They were.

New buildings were springing up all over Riyadh and Jiddah, but they were jerry-buildings rushed up by foreign contractors for huge profits. Several of Ibn Saud's sons had gone to San Francisco for the inauguration of the United Nations at the end of World War II, and had thrilled large numbers of the local

151

F

females with their doe-brown eyes and romantic Arab robes. They brought many of them back with them, and picked up others in London and Paris, and they turned out to be high-priced whores who quickly got their fingers into the princely incomes. According to Harry Philby, one of them married a prince and shortly afterwards spent a million dollars on a European shopping spree.

The princes brought back from the inaugural meeting none of the high principles which the United Nations declared as its ideal in those days. One of the king's sons, when asked what had impressed him most in America, cited the miniature mermaid floating in the trick tank in the entrance hall of a New York nightclub. They loaded down ships with Cadillacs and spares, with gallons of perfume, and cases of drink. They brought in hundreds of thousands of coloured electric lights with which to festoon their houses and gardens. When the cars broke down, they left them in the desert and sent for more. They hired French chefs, but found the food insipid and threw it away.

As Riyadh increased in prosperity, more and more waste food was thrown into the streets [wrote van der Meulen]. The army of mongrels grew in numbers and as a nuisance. Their nocturnal fights made life in town unbearable so it was decided to get rid of these pests. Since no Muslim will simply kill an animal because it barks, the dogs were not shot or poisoned, but in the desert at some distance from the town high, square mud enclosures were built. At the entrance of each enclosure was posted a town official who paid three Saudi rials for every dog delivered to him. When we visited these pounds the guards told us that between four and six thousand dogs had passed in. I looked through the crack between the gate and saw a host of fly-pestered dogs. For long I could not take my eyes off the suffering animals . . . When I asked one of the government officials about the dogs he told me that they were given water and old dates, and would doubtless die when their time came.[3]

But, as Ibn Saud was discovering for himself, death does not always come quickly or easily, and he had to sit by, enfeebled and helpless, and watch the country he had single-handedly constructed going to pieces before him. Scandals involving the princes became the talk of the kingdom. A British consul was shot dead by a drunken prince at a party; two European women and a number of Saudi men died of poisoned alcohol at another princely celebration. The king ordained death for the culprits but the sentences in each case were never carried out. He blamed the Americans of Aramco for bringing sinful alcohol into the country and ordered a rigid ban on its use, even in the privacy of the oil company's cantonments; but what he did not know, and what no one dared to tell him, was that the princes themselves were the importers of liquor and made money on the side by selling it.

And for all Saudi Arabia's vast income from oil, money was scarce and the bulk of the people existed at starvation level. There were schools – but they had been built by Aramco. There were hospitals and good medical services – but they were provided and maintained by Aramco. The oil company had even acceded to one of the king's last whims and built him a railway, from Dammam to Riyadh across the desert. But where the oil company had no wells, no offices or cantonments, there was poverty and disease. The money that might have gone into public welfare was squandered by the court under Ibn Saud's single, lack-lustre eye.

The US government has been blamed for having allowed Aramco to pour so much money into Ibn Saud's coffers and doing so little to see that it was wisely spent.

It was also wrong that the money given to Ibn Saud was given with hardly a word of reliable advice on the spending of it [writes David Howarth in his biography of the king]. Aramco could not be blamed for that; it was only an oil company and its only reason for being in Arabia, or existing, was frankly to make a profit . . . [It] would hardly have been expected to act like a mandatory power, and that was what was needed. Advice was a job for a government [4]

And van der Meulen writes:

Are the Americans to be blamed for the scale on which the millions poured into one of the poorest countries in the world were swept out again into private hands? Could they not have done something about it? [5]

The answer US Arabists give to that is that advice was offered from time to time, and those who gave it were rebuffed. Aramco made it known that they would have preferred to pay Saudi Arabia's share of the oil revenues into a government fund rather than the king's private purse, and were crisply informed by Abdullah Sulaiman, who controlled the royal revenues, that the king would be so offended by any attempt to do so that he might cancel the concession. These were the postwar years when Britain, France and Holland were giving up their colonial empires. It is easy to imagine what their comments would have been had a US government stepped into an independent Arab state and announced that it would henceforth be managing its budget.

Aramco itself was in a difficult position. Most of its directors were oilmen who had worked their way up from the rigs. They had developed a love of the desert and a certain paternal affection for the Saudis with whom they worked, and they were proud of their achievements in training, building and improving. To an overwhelming extent, any evidences of enlightened building along the Saudi shores of the Persian Gulf were their creation. They had not just drilled wells but built roads, a railway, and whole towns. They had dug irrigation ditches and made the desert sprout vegetables and fruit. They had brought education and medicine to a backward people, and wherever they worked the standards of literacy and health among the Saudis was higher than where the king's writ ran alone.

But they were a company with two loyalties. On the one hand they had to satisfy the demands of their four parent corporations (Standard of California, Standard of New Jersey, Texaco and Socony Vacuum – now called Mobil) for highly profitable returns. On the other, they had to keep the king supplied with more and more money. There was little they could do about the fact that the money was promptly wasted in foolish extravagances or poured into the coffers of the swindlers around the court. What Ibn Saud's ministers wanted was not advice but more revenues, and the more they were paid the more they spent, and the deeper the kingdom slid into debt. By 1950, after J. Paul Getty had paid $10 million for the concession in the Neutral Zone, the king's ministers became exigent.

The Saudi Government had just entered into the contract [with Getty], [said F. A. Davies, president of Aramco*], and the terms were much better than ours. Our concession had greatly increased in value. We had developed a big reserve out there. They asked us as early as 1948, 'Isn't there some way in which we can get a greater take?'[6]

* Not to be confused with R. K. Davies of Aminoil.

Now they began demanding so much money that Aramco's lawyers in the US believed that in order to satisfy them the whole nature of the company's concession might have to be altered in order to provide it, and that was something they wished to avoid at all costs. Moreover, the payments, if made out of Aramco's revenues, would have wiped out a large proportion of the company's profits, and the parent companies in the US were in no mood to accept that. 'They had shareholders to satisfy,' Davies said.

But how could a way be found to satisfy the greedy demands of the king's courtiers without dealing a drastic blow to Aramco's profitability?

The legal staff of Aramco knew that in 1949 the company had paid $38,000,000 to the Saudi government. But in the same period the company had given $43,000,000 to the US government in income tax. The fact that the Saudis were receiving less from their own mineral wealth than a government thousands of miles away was something which Aramco officials had kept discreetly quiet until now. But suddenly, on the advice of the company's lawyers, the figures were leaked to the Saudis. The result was that (in the words of F. A. Davies) 'they weren't a darn bit happy about it'.[7]

Soon they were asking the question that anyone with Aramco's profitability at heart would have wanted them to ask: 'Isn't there some way in which the income tax you pay to the United States can be diverted to us in whole or in part?'[8]

It was at this point that the company suggested that the Saudi government consult the US Treasury. Aramco had already discussed its own problems with Saul McGhee, of the Department, who according to Davies, 'appreciated our difficulties', and the net result had been the dispatch to Jiddah of a Treasury Department official, George A. Eddy, who had conferred with Saudi officials about their money problems. When asked a direct question by a Saudi official as to how more money might be raised from 'foreign firms', Eddy had first consulted the US ambassador in Jiddah as to whether he might answer the question, and then, given permission, had pointed out that several methods were available. One of them was (in the case of an oil firm like Aramco) to demand an increase in royalties on oil produced; the other was to institute an income tax system and get more money from the company by direct tax. Eddy added: 'I did explain to him [the Saudi official] the difference of the effect on the company of a royalty and an income tax.'[9]

By this he meant that if the Saudi government simply increased the amount of royalty it was receiving from Aramco per barrel of oil, it would have a direct (and damaging) effect upon the profits of the company. If, however, the Saudi government were to start an income tax system, any money paid to them by Aramco as such a tax could (under US law) be deducted from the amount of tax the company was liable for in the United States.

Eddy went back to the United States. He was followed to Saudi Arabia by a Washington lawyer, John F. Greaney, who subsequently drafted an Income Tax Law for the Kingdom of Saudi Arabia which was instituted by royal decree on 26 December 1950. It came a day late, but it was still a munificent Christmas present for both the Saudi government and Aramco. Under the last year (1950) of the old system of payment by royalty only, Aramco paid $56,700,000 to the Saudi government. In the first year of the new system of royalty, lease and income tax combined, the government got $110,000,000.

The nicest thing about the new system from Aramco's point of view was that it didn't cost them a penny. They simply wrote off their Saudi Arabian taxes against their liabilities for US tax.

Senator McHugh: 'The net result is that the United States Government now gets nothing in the way of income tax. Is that correct?'

F. A. Davies: 'They haven't received any taxes from us for two or three years, I guess.'[10]

Soon governments with oil concessionaires all over the Middle East were adopting the system. It was called the fifty-fifty system, since it gave the government of the producing country approximately half of the oil operating company's earnings, and to begin with, everyone was more than satisfied with it.

Except, of course, the government of the United States, which now began to lose something like $100,000,000 in tax revenues every year from American oil companies operating in the Middle East.

On 9 November 1953, King Abdul Aziz Ibn Saud, Lord of Arabia, died in his sleep in his palace at Taif, a hill station forty miles east of Mecca. As Harry St John Philby was later to remark, there were few people around – save himself and one or two other European Arabists – who mourned his death. He had been one of the great heroic figures of Oriental history, and he had carved the new country of Saudi Arabia out of the desert with his own sword. But the man who had led the tribes to victory in battle had allowed them to relapse into squalor thereafter, and there were few ordinary Saudis who could be said to have benefited from his reign. They had more reason to thank Aramco than the king for any progress they may have experienced.

As for his sons and his venal ministers, they had long since turned away from him to more profitable sources of influence.

A hush fell on his palaces long before his death [wrote his biographer]. The throngs of visitors and servants and supplicants diminished, dwindled, lost their urgent air. An air of stealthy secrecy replaced it because there were so many things that had to be hidden from him.[11]

His body was taken by night to Riyadh and buried quietly, with no ceremony, according to the custom of Wahhabi Muslims.

He was succeeded on the throne of Saudi Arabia by his son, King Saud. The new monarch made a speech to his people on the day of his elevation in which he said: 'My father's reign may be famous for all its conquests and its cohesion of the country. My reign will be remembered for what I do for my people in the way of their welfare, their education and their health.'

In his testimony before a Senate Committee in 1957, the president of Aramco, F. A. Davies, said:

What he [the new king] has been doing in the past few years bears that statement out very much. He's building schools, he's building homes, he's building mobile clinics, he's building clinics that aren't mobile all over the country. He's building roads, water-wells, developing agriculture, he's developing the mosques. Too much is said here in the magazines about the gold-plated Cadillacs. I've not seen any gold-plated Cadillacs. I've seen Cadillacs but I've seen lots of Cadillacs in Texas ... I want here to subscribe very heartily to the programme that His Majesty is putting over in his country to scotch ideas that that there is nothing but waste there. There are many good things being done there and each semester, each year you can see greater increases.[12]

At the moment the Aramco president was speaking – and well must he have known it – the new king had surpassed anything his father had achieved by running Saudi Arabia into debts totalling just under five hundred million dollars. He had a harem that far outstripped in numbers that of Ibn Saud, even though he may not have serviced it so efficiently. He had a vast army of hangers-on at court who were allowed to indulge in the wildest extravagances.

The new king, to whom Aramco's president had paid such eloquent tribute, was profligate, decadent, and an unmitigated disaster for his country.

15 · The Man in the Pyjama Suit

Early in May 1950 the remains of Iran's Shah-in-Shah, Reza Khan, who had been toppled from his throne by the British and the Russians during World War II, arrived back in his homeland from Johannesburg, South Africa, where he had died in exile. Though he had done much during his reign to regenerate Iran, the old shah had never really been popular with his people, but his reburial was made the occasion for elaborate ceremonials. Detachments of Pakistani, Iraqi and Turkish troops marched with their bands past the decorated catafalque in the wake of a great procession of men, guns and tanks of the Iranian Army. The streets of Teheran were packed with spectators staring in superstitious awe at the imperial coffin as it was trundled on a gun-carriage, drawn by four black horses to its final resting place on the outskirts of the city. For the most part the crowd waited and watched in silence, but when the British representative was sighted in the delegation of diplomats marching behind the coffin, voices rose and fists were shaken, and there were cries of:

'Our shah has come back! Now give us back our oil!'

Among the corps of marching members of the Majlis, the Iranian parliament, there shuffled by a long thin, bent old man with a haggard yellow face and dripping nose which he did not bother to wipe. Great cheers rose from the throng. This time the crowd began crying a name:

'Mossadeq! Mossadeq! Mossadeq!'

They were hailing the leader of a newly-formed coalition party, the National Front, and Doctor Mohammad Mossadeq was popular with the crowd because the main plank in his political platform was to wrest control from the British of Iran's petroleum resources.

Despite the fact that those resources were now among the richest in the world, Iran in 1950 was still a backward country, racked by disease and corruption, manipulated by squalid politicians and landowners, and ruled by a shah (son of the old shah, now being buried) who was weak, vacillating, uncertain and far different from the arrogant and self-confident monarch he was to become by the 1970s. Only in the south-west, where Anglo-Iranian exploited the nation's oil, could it be said that rewarding work and a decent standard of living had been given to the populace. In Abadan, in Fields, in Ahwaz and Masjid-i-Sulaiman and all the other communities which had grown up around the wellheads, the refineries and the tanker ports, the British company had brought civilization: built houses and hospitals and schools, drains for sanitation, roads for communication, clubs and cinemas and playing fields. As a commission from the International Labour Office reported after a visit in 1950 to Anglo-Iranian's operations:

The observer cannot fail to be impressed by the vast numbers of modern houses and amenities which the Company has been able to provide in a comparatively short time, in spite of exceptionally unfavourable circumstances.[1]

But the British were discovering, not for the first time in their colonial history, that these bounties did not make them loved by the people to whom they were given. Nationalists know no gratitude, and the fact that it was British know-how, British guts and persistence, and British money which had found the oil in the first place cut no ice with them. Why should it? The British had already earned an enormous return from their outlay in muscle, brains and finance. To the pioneers all honour and a just reward, but why go on pouring out millions in profit to men who had come in only after the spadework was done, and to shareholders for whom Iran was only a name in an annual report? The nationalist politicians lusted after the vast fortunes which Anglo-Iranian was making. The workers, especially trained fieldmen and office staff, yearned for the higher echelons which, they believed, were only denied them because the British had all the important jobs. To the Iranian people it seemed that by taking over Anglo-Iranian all their problems would be solved, there would be jobs galore, money would be plentiful, and their sense of humiliation would be wiped out because Iranian resources would be back in Iranian hands.

So they listened avidly to any politician who, as Mossadeq did, preached nationalization of Anglo-Iranian as the universal panacea. And he did not hesitate to call the British cheats for the way in which they were milking the company's earnings, at Iran's expense. He had a point. If Anglo-Iranian was not exactly cheating its host-country, it was certainly paying bargain prices for its enormous harvests of oil. The concession agreement which it had made with the old shah was based on a royalty payment for every barrel of oil extracted, and in 1950 the amount paid to the Iranian government was £16,000,000. In addition to this, the company sold petroleum to the Iranian government at exceptionally low prices and the government in turn sold it to the people at high prices, earning another £7,000,000. And through Anglo-Iranian's intercession with the British government, the Iranians were allowed to manipulate the exchange rate between sterling and Iranian rials, and thus earn themselves another £5,000,000. In this and other minor ways, the Iranians received from the British about £32,000,000, and it constituted practically half of the Iranian budget.

But (and this was what the Iranians and their friends most resented) Anglo-Iranian was earning nearly five times as much from the Iranian oilfields, and paying in taxes to the British government over £40,000,000 – £8,000,000 more than Iran was getting from its own oil.

This was a glaring inequity, and the British were later to accuse the then US ambassador to Iran, Dr Henry Francis Grady, of 'gross disloyalty' towards them by pointing it out both to the young shah and to Mossadeq. They charged that the American diplomat was trying to 'sabotage' them in order to get them expelled from Iran, and US oil experts substituted. In fact, unlike some US advisers who came later, Dr Grady was no particular friend of any oil company, either British or American. He was, on the other hand, keenly interested in helping the Iranians to find more money. With the help of US consultants, the Iranians had been persuaded to draw up a

Seven Year Development Plan and had been promised US loans with which to implement it. The more money they could provide themselves the less would be the burden on the American taxpayer. Grady stressed in his messages to Washington that Mossadeq might seem a fiery and eccentric fanatic – he had a habit of giving press conferences lolling on couches, dressed in pyjamas – but he expressed the sentiments of the people.

Aware of the way things were running, Anglo-Iranian decided that the moment had come to make a gesture. A proposal was made to increase royalty payments to such an extent that they would have brought Iran's income from oil in 1950 up to £40,000,000. But the prime minister, who presented it to the Majlis, read it out in a statement that was so patently translated directly from English – and had almost certainly been written by the British embassy and Anglo-Iranian officials in collaboration – that Mossadeq's supporters accused the premier of being a British puppet and shouted him down. The proposal was finally rejected the following year, and Anglo-Iranian, by this time beginning to be alarmed, proposed that Aramco's example in Saudi Arabia be followed, and a fifty-fifty deal be consummated between the company and the government.

But the British had seen the writing on the wall too late, and by the time they got around to making this proposal the moment had passed. Everyone in Iran was hell-bent on nationalization. Or rather, nearly everyone. The young shah, for instance, was in a dither. He sat insecurely on his peacock throne. He could not decide what to do about nationalization. His instinct was to support it, but his political advisers kept telling him that it would bankrupt Iran and cost him his throne. Finally, he sent instructions to his ambassador in Paris, Ali Soheily, to go to Lisbon and ask Calouste Sarkis Gulbenkian whether he should oppose it or espouse it.* What about Anglo-Iranian's latest offer? Should he accept it?

Mr Five Per Cent told Soheily to tell the Shah to accept Anglo-Persian's offer [said Nubar, later]. Father pointed out that it was the best ever proposed by an oil company in the Middle East and ended

---

* Both Gulbenkian and his son, Nubar, held Iranian diplomatic passports as 'honorary attachés' of the Iranian embassies in the countries in which they resided.

by declaring bluntly, 'Our country is not yet qualified to take over the oil industry.' Soheily found this opinion very embarrassing. The British offered to put a special plane at the disposal of Father and me (and our suite of five persons) so as we could fly to Teheran to give the Shah our views personally. But Mr Five Per Cent refused to move. Soheily himself had to report back the unpleasant news.[2]

On 19 February 1951, General Ali Razmara, announced in the Majlis that according to the oil experts whom he had consulted nationalization was completely impracticable. Once more the tone and manner indicated that the statement had been translated from the English. Razmara was hooted down, and shouts of 'traitor!' and 'bootlicker!' were hurled at him. Shortly afterwards, as he knelt to pray in the mosque, he was shot dead by an assassin.

On 29 April 1951, Doctor Mohammad Mossadeq swept into power as head of the National Front government, and immediately presented a bill before the Majlis for the nationalization of the Anglo-Iranian Oil Company. It was acclaimed unanimously. Next day the bill was signed by the shah, and passed into law on 31 April. Iranian oilfields were now Iranian. It was the end of Britain's control, after a monopolistic reign which had lasted nearly fifty years and fuelled the British Empire in war and peace. Now the monopoly was no more.

Not that the British believed it, at first. 'They can't do this to us,' said the British ambassador, Mr Francis Shepherd, in a statement to press reporters in Teheran. MOSSIE GRABS BRITAIN'S OIL – BUT NAVY TO THE RESCUE, headlined the *Daily Express*. For, as if this were 1905 instead of 1951, the British were planning to send a gunboat to get back what the Iranians had taken away.

The events of the next few weeks, looked back upon from the more realistic standpoint of the 1970s, have a true-blue tinge to them that turned what was after all a quarrel between a landlord and his tenant over the terms of his lease into the last act of a farce about imperialism. Though the Labour party was still in power in Britain at the time, which might have predicated a sympathetic attitude towards another nation's nationalization plans,* British reaction to the takeover of Anglo-Iranian

* The Labour government, after all, had nationalized several industries of its own in recent years.

was everything that a stout-hearted Tory could have asked for. An echo of its imperial tones are to be found even today in some British accounts of the crisis. The British oil historian, Stephen Hemsley Longrigg, for instance, writes:

The ratification of this naive and totally inadequate [nationalization] law—henceforward to be the inviolable basis of all discussion and negotiation—was followed by an enthusiastic popular and political campaign of abuse and misrepresentation, directed against the Company. . . Forces of dangerous, indeed of bloodthirsty, fanaticism were allied to and largely dictated the moves of the nationalizers, who became prisoners of their own law and of their broadcast promises of wealth and happiness for all; and the contributions of certain American diplomats and consultants, who professed to see the situation as one of Anglo-Iranian 'colonization' or mere reaction or as a legacy of unspecified Anglo-Iranian misdoings in the past, encouraged even responsible Persians to expect American approval.[3]

Soon both sides were striking attitudes which everyone had imagined had gone out of fashion with the death of Queen Victoria. For the British, the wogs were on the rampage. For the Iranians, a war of liberation had begun against the colonialists. What started out as reasonable discussion soon deteriorated into vituperation, threats and violence. Britain submitted the nationalization law to the International Court at the Hague, but when the court (which included a British member) finally pronounced, the result was deadlock. The court refused to adjudge what they considered to be a 'domestic' affair, and told the contenders to go back and talk it over. It was no longer the question of a quarrel between a British-owned business and a foreign government. The British authorities had taken over, and now all future moves were decided from No. 10 Downing Street. To begin with, Anglo-Iranian was told to halt all royalty payments. The arrangement by which the Teheran government could change rials into sterling at bargain rates was ended forthwith, producing an immediate famine in foreign currency for Doctor Mossadeq. Meanwhile, back in the oilfields, the three thousand odd British technicians who worked for Anglo-Iranian had been offered new contracts by Teheran to go on working for the new national company. Unanimously, they refused. Tanker captains arriving at the oil ports to load up with crude were told by Iranian officials to sign receipts acknowledging Iranian ownership: they too refused and sailed away with empty ships. Soon one field after another began to close down as the pumping stations and pipelines ceased to function.

This was a signal for riots to break out. Anglo-Iranian's chief in Teheran had his house sacked by mobs. There were mass demonstrations in Abadan and other main centres of the oil-fields. Mossadeq, lying torpid on his bed in his pyjama suit, or dashing suddenly to New York to address the United Nations, called the riots signs of 'the natural reaction of the people to generations of looting by the British'. The British replied by evacuating all British women and children, and sending ministerial messages of encouragement to the men left behind in the oilfields to stand by and keep their chins up. HOLD FIRM, BRITAIN TELLS OILMEN, WE'LL STAND BY YOU, MINISTER PROMISES, was a headline in the *Daily Mail*.

At which point the situation escalated again when the British government dispatched a parachute regiment to Cyprus and ordered it to stand by for further instructions. British army units were alerted in Iraq, just across the border from the Iranian oilfields. And suddenly the Royal Navy cruiser H.M.S. *Mauritius* hove-to off Abadan and anchored in mid-river. It looked as if a landing and a military occupation of the oilfields was imminent.

At the last moment, however, that disastrous project was abandoned, luckily for Britain. Otherwise the defeat she was to suffer at Suez five years later could well have happened that much sooner. For the Iranian Army was ready and itching to fight, and had the will of the people behind it. Britain had laid on no landing plan, preferring an ad hoc operation. Moreover, her plans of communication could have been seriously com-promised throughout the Middle East, for every Arab nation without exception approved of Mossadeq's actions. More important than all these drawbacks, however, were the two biggest threats of all to the success of any military operation. One was threateningly negative: the United States government let it quietly be known in Downing Street that under no cir-cumstances would America give any aid or support to a British punitive action. The other was threateningly positive: the Russians began making dispositions on their frontiers with Iran, and there were indications that in the event of a British landing in the south the Red Army would march into the country and establish a puppet Soviet regime in Teheran.

Faced with these horrid facts of postwar life, the Labour government wisely decided to back down and stop acting like a gunboat administration. Instead of reaching for their guns, they started to use their heads – and suddenly realized that they had the strongest weapon of all on their side. Time.

In October 1951, the last members of the British staff in the Iranian oilfields were evacuated aboard H.M.S. *Mauritius*. They went out like true-blue Britons.

On the morning of October 4, 1951 [writes the official historian of Anglo-Iranian] the party assembled before the Gymkhana Club, the centre of so many of the lighter moments of their life in Persia, to embark for Basra in the British cruiser Mauritius. Some had their dogs, though most had had to be destroyed; others carried tennis rackets and golf clubs; the hospital nurses and the indomitable Mrs Flavell who ran the guest house and three days previously had intimidated a Persian tank commander with her parasol for driving over her lawn, were among the party, and the Rev. Tyrie had come sadly from locking up in the little church the records of those who had been born, baptised, or had died in Abadan . . . The ship's band, 'correct' to the end, struck up the Persian national anthem and the launches began their shuttle service . . . The cruiser Mauritius steamed slowly away up the river with the band playing, the assembled company lining the rails and roaring in unison the less printable version of 'Colonel Bogey'. Next day Ross and Mason [the two senior officials] drove away. The greatest single overseas enterprise in British commerce had ground to a standstill.[4]

Behind them they left pipelines through which the pumping stations were no longer pumping oil, closed-down refineries, tank-farms spilling over with crude that no tankers were moving, and 70,000 Iranian employees still on the payroll for a salary bill of £1,600,000 a month. The responsibility for paying them was now that of Doctor Mossadeq's regime, but he certainly did not have the money. His government had other obligations, no less pressing, which he hadn't a hope of fulfilling.

For Britain too the situation threatened to deal an almost fatal blow to the economy. Anglo-Iranian had always been one of the United Kingdom's greatest earners of foreign currency, and the markets it had served were now lost. The company had also supplied the bulk of Britain's petroleum needs, and now these would have to be replaced from other sources, almost all 'hard' currency sources which would cost the British treasury £40,000,000 a month at a time when dollar reserves were dangerously low.

It was a desperate situation, and it explains why even a liberal-oriented Labour government had, for a time, contemplated a punitive expedition to occupy the Iranian oilfields by force. The chaotic economic situation created by the nationalization did nothing to help them during the general election which took place in Britain at the height of the crisis, and the result was that Winston Churchill and the Tory party were voted back into power. Churchill's first action was to reach for the telephone and talk to the White House, and ask for US aid in resolving the situation.

On 15 July 1951, Mr Averill Harriman arrived in Teheran on a special mission for President Truman. He came with preconceived ideas about the Iranian situation, for the US ambassador, Doctor Grady, had made his views quite clear in his reports, stressing the fact that Mossadeq enjoyed the support of most Iranians, and castigating the British for the arrogance of their local diplomats and the stubbornness of their oilmen. Harriman therefore arrived as the envoy of an administration which, while taking no sides in the dispute, sympathized with Iranian aspirations, and plainly expected to be welcomed as a friend as well as a go-between. Instead what he got was a flea in his ear.

For Doctor Grady's reports were now out of date, and what had started out as a genuine popular movement of protest against foreign exploitation (real or imagined) had now been taken over by the ideologues. The Iranian Communist party, Tudeh, which was controlled and financed from Moscow, had now thrown its weight, and its money, behind Mossadeq, and provided him with a street army of well-trained thugs. The Tudeh party was not in favour of oil nationalization, for it supported Russian claims to oil concessions in northern Iran, and now hoped to engineer a Russian control of the fields from which the British had been driven; so it directed its campaign to attacks on 'colonialism' and 'dollar imperialism', which brought the United States in alongside Britain as the enemy to be attacked. Harriman, all prepared to smile his friendly smile and wave an acknowledging hand to the cheering crowds he expected to meet on his way in from the airport, instead encountered an ugly scene with a column of Tudeh demonstrators armed with sticks, and screaming anti-American slogans.

He was somewhat taken aback at his first interview with Doctor Mossadeq when the Iranian leader, in the midst of an outburst against the British, suddenly dissolved into tears. This unexpected emotionalism must have affected Harriman's sense of judgement, for he emerged from the meeting convinced that providing Britain expressed herself ready to recognize Iran's right to nationalize its oil resources all would be well and reasonable discussions could begin. On the strength of this, the British government rushed an envoy out to Teheran, only to be confronted by a demand from Dr Mossadeq that they pay $140,000,000 into his Treasury immediately – and no money, no talks. Harriman expressed himself both surprised and disappointed.

Harriman returned to Washington a man whose ideas had changed. By the time Truman's administration had given place to that of President Eisenhower, there was no longer any question of America standing aside and letting the British get out of their own mess. Churchill and Eisenhower were old friends and spoke the same language. The Republican administration had always been responsive to the Oil Lobby, and the oilmen, though far from being discomfited by their British rivals' difficulties, saw the peril ahead if a Middle Eastern nation were allowed to get away with nationalization. What Iran could do today Saudi Arabia might do tomorrow.

So far as the oil-supply situation was concerned, time was on Britain's side. So far as Mossadeq was concerned, it was running out. The old man's trouble was that he had never visited the Iranian oilfields and he knew nothing at all about the situation there. He had never bothered to ask anyone on the spot whether – once the British left – there was a sufficiently trained cadre of specialists capable of running the fields. There was not. The British had schooled several thousand Iranians to take over the lower technical rungs of the petroleum ladder, but there was no one capable of climbing to the top and taking over.

Nor had Mossadeq realized that there is more to the oil

business than getting it to the surface and refining it. It must then be sold. And since no Iranians in the Anglo-Iranian Company had held important positions in the sales department, no one had explained to them the facts of life in the international oil sales market. There were 12,000,000 barrels of refined petroleum lying in the brimming tanks at Abadan, and the Iranian leader demanded that it be sold at once to earn desperately-needed foreign currency to meet his budget.

'No one will buy it,' the new owners sadly told him.

'Then sell it at half price – at a quarter price,' said Mossadeq.

But still no one would buy. Anglo-Iranian had made a declaration immediately after withdrawing from Abadan asserting its proprietorial rights over any oil coming from their former fields, and threatening action for compensation against anyone in any country who tried to sell or buy Iranian oil. The first to try it were the Italians. Italy had no oilfields of its own. It had always relied entirely on Anglo-Iranian for its fuel supplies, and when nationalization took place the Italians indicated – Anglo-Iranian's threats notwithstanding – that they would go on buying from Abadan. But then the ebullient head of the national Italian petroleum company, AGIP, a one-man dynamo named Enrico Mattei, flew to London for conferences with Lord Strathalmond, chairman of Anglo-Iranian. He emerged from the meetings declaring that he had not previously taken the British threats of law action seriously, but that now he did, and AGIP would not be buying Iranian nationalized oil.

Everyone presumed that Signor Mattei had been offered powerful inducements and substantial future rewards for having changed his mind.

A small Italian company, SUPOR, did take the risk and loaded two tankers at Abadan. One had a writ tacked to its funnel when it arrived in Aden and the other when it reached Naples. A Japanese tanker which took a load to Osaka was also sued by Anglo-Iranian in Japan.

Nonetheless, these law cases would take months to come to trial, and if enough companies had been willing to take the plunge, no legal threats from the British company could have stopped the boycott from being broken. But the international oil market did not want the boycott broken. Once Iranian oil was phased out, they discovered that they were doing even better without it. Saudi Arabia, Kuwait and Qatar oilfields stepped up production and turned 1952 and 1953 into bonanza years for the big American companies and for the Anglo-American-Dutch-French combine, IPC. By the end of 1953 world oil production had risen to 637 million tons (compared with 535 million in 1950). In comparison, Iran (which had exported 54,000,000 tons of oil through Anglo-Iranian in 1950–51) sold a total of only 132,000 tons to foreign buyers under Mossadeq in 1952–3.

Belatedly made aware of this miscalculation, Mossadeq searched around for someone – anyone – who could sell oil for him. He tried Enrico Mattei – but by this time Mattei had seen Lord Strathalmond. Then he rushed an emissary to London to see Nubar Gulbenkian. The Gulbenkians had strongly advised the Iranians not to nationalize Anglo-Iranian, and for their pains had been deprived of their Iranian diplomatic status the moment Mossadeq had come to power. Now Mossadeq offered Nubar Gulbenkian plenipotentiary powers and 'the gratitude of the Iranian people' if he would take over as sales director or adviser of the nationalized company. Nubar refused ('though it would have been nice to put the CD plate back on my car') and went to Lisbon to let his father know what was happening.

'I told them so,' the old Armenian said, happy that yet another of his predictions about the oil business had come true.

By now diplomatic relations between Britain and Iran had been broken off – on Mossadeq's orders. The United Kingdom complained to the United Nations, and also used America's good offices to make a last offer to patch up the quarrel. In February 1953, new proposals were sent to Teheran and the US government indicated that they had her full approval. In return for a settlement, both Britain and the US would:

1 Recognize Iran's right to control her own oil industry and her own oil policies;

2 Expect compensation to be paid to Anglo-Iranian for this change in status;

3 Arrange for Iran to have a full opportunity to enter into arrangements for selling oil in substantial quantities at competitive prices on the world markets; and

4 Lend Iran sufficient funds (to be repaid in oil) to enable the government to meet her immediate financial problems.

The proposals were rejected in a fiery speech from the

Iranian leader in which he launched into his most virulent attack yet on the British oil company, angrily refusing to contemplate paying compensation.

'What the Anglo-Iranian Company did in Southern Iran,' he cried, 'was sheer looting, not business!'

He was determined not to have the British back, but he was also now direly and disastrously in debt. Oil workers had been put to labour on the roads, but the government was running out of money to buy supplies to keep them working. A US loan of $23,000,000 had been enough to pay for shipments of food, but its arrival was said to have resulted from a personal appeal from the shah to the new US ambassador, Mr Loy Henderson, and since Mossadeq was now quietly working for the shah's overthrow he was anxious that the monarch should get no credit for any further aid.

So when, on 23 May 1953, the Iranian leader asked the US government for further loans, he made it clear that this time they would be on his terms. He sent a personal letter to President Dwight Eisenhower soliciting the financial aid he needed, and he added what amounted to an ultimatum in the final paragraph. Calculating that there was nothing more likely to speed up action in Washington than a threat of Soviet intervention, he let it be known that if the United States could not provide him with the help needed, then he would be forced to look elsewhere for aid. His ambassador in Washington underlined what that meant. The Soviet government had already made it known in Teheran that it was ready to help, and was prepared to sign a defence pact as well as an economic agreement with the Mossadeq regime.

It was a threat just as likely to frighten the Americans as the British – the threat of Red Russian penetration into Iran and down to the waters of the Persian Gulf was something neither nation liked to think about. Mossadeq figured that it would produce precipitate action.

It did. It brought the CIA into the crisis.

16 · The Flight of the Shah

Teheran in 1953 was so full of American and Russian spies that a local wit once suggested that they ought to share apartments, to save themselves time and money in keeping each other under surveillance. There was a residue left in the Iranian capital of the Polish and Balkan refugees who had been washed up there during World War II, and most of them now worked as *filles de la maison*, musicians or waiters in the plethora of cabarets sprinkled around the city. Many of them were on the payroll of one or other of the two embassies (several of them on both) and found a ready market for the gossip they picked up during their nocturnal preoccupations. The Russians had also infiltrated agents into most Iranian government departments and business offices in the city, and the bazaars swarmed with characters who carried documents to prove that they came from Iranian Azerbaijan when they came, in reality, from Soviet Azerbaijan, just across the northern frontier. A number of these, sent down from the oil-fields in Baku, were also operating in Abadan and Ahwaz, trying to get order into the chaos of the nationalized petroleum operations.

The Americans had a military mission attached to the Iranian Army which Mossadeq, not wanting to break completely yet with the United States, had reluctantly refrained from expelling, and its members were in close touch with their Iranian confrères. There were also strange Americans named 'Jake' and 'Red' and 'Uncle Ami' (it would serve no purpose to give their full names) who made regular appearances around the cabarets and bars, speaking purest Bronx, Brooklynese or Panhandle Texan, but turning out to command an even more fluent flow of Kurdish and the dialects of Kermanshah, Khorammbad and Azerbaijan, learned at the knees of their immigrant parents.

They came in to report and quench their thirst for whisky and female company, and then disappeared again for weeks on end into the blue.

But the section of the Iranian administration into which American agents had infiltrated with most success was the Iranian police force. From 1942 until 1948, the force had been under the command of an American named Brigadier-General Norman H. Schwarzkopf, who had orders to recruit and reorganize it. General Schwarzkopf, whose name will be familiar to students of the Lindbergh Baby kidnapping case,* had now gone on to bigger things, and had a roving commission for the CIA, under Allen Dulles. Both of them had been wartime members of the OSS.

When he departed from Teheran, Schwarzkopf had left several old friends behind in the Iranian police who could be relied upon in an emergency. They had, in fact, already proved their efficiency in the spring of 1953. Aware of the pro-American elements in the police ranks, Mossadeq had appointed a new police chief and given him orders to conduct a ruthless purge. The new chief was General Ashfar-Tus, and there was no chance that he would take his instructions lightly, for he was a savage and ambitious man. His character can be judged by the fact that a fellow general, with a reputation for being one of the sternest disciplinarians in the Iranian Army, had sacked him 'because his behaviour with the cadets was so harsh and brutal'. Afterwards he had been given charge of royal estates in the north and there 'treated the peasants with brutality and even cruelty'.[1]

General Ashfar-Tus had one fault. He boasted. On the night of his arrival at police headquarters on 19 April 1953, he announced to a number of his subordinates that he had a list of the names of every American spy in the force, and promised that they would all be dead or behind bars by the end of the week. But next morning he did not turn up to make good his threat. He had been kidnapped, and his corpse, shot through the head, was discovered in the foothills of the Elburz Mountains some time later.

Mohammad Mossadeq had sent his letter asking for a loan to President Eisenhower on 23 May 1953, but a month passed by

* He was in charge of New Jersey State Police in 1932 and headed the hunt for the kidnapper.

and no reply came from the White House. One can only presume that the president had been advised to delay and to keep the Iranian leader in suspense about whether he would receive the loan he needed until certain preparations had been made. Certainly, there was a sudden flurry of activity among CIA agents both inside and outside Iran during the hiatus. Mr Allen Dulles, the CIA chief, flew to Switzerland where, it was said, he would be joining his wife for a short holiday. A few days after his arrival, who should turn up to join him but Loy Henderson, US ambassador to Iran. And a week after that, Princess Ashraf, the shah's sister, and one of the most forthright characters (male or female) in the kingdom, also turned up in Switzerland and had several discreet meetings with the two Americans.

It was not until 29 June 1953, that President Eisenhower at last got around to answering Mossadeq's letter, and the Iranian leader was jolted by the unequivocal nature of the president's reply. It too contained an ultimatum. Either Mossadeq agreed to an immediate opening of talks on the future of the oilfields, or there would be no loan. The old man gave a bedside press conference during which, either quivering with rage, or shaking with sobs, he swore that he would never give in to the imperialists who were trying to humiliate the Iranian people. Strike calls were answered by Tudeh parades in the capital and the oilfields, hailing Mossadeq and the Soviet Union, and promising to blow the oilwells sky high should British or American troops set foot on Iranian soil. The army and the police force, on American advice, stayed in their barracks and left the streets in possession of the mobs.

In the midst of the turmoil, who should arrive in Teheran but Brigadier-General Norman H. Schwarzkopf, armed with a diplomatic passport and a couple of large bags. As soon as the news was known that General Schwarzkopf was in the capital, cries of fury and alarm rose up in the Tudeh- and Russian-controlled newspapers, and members of the Soviet Economic Commission, which was in Teheran for discussions with the Iranians, formally protested the presence of 'this notorious agent of American Intelligence'. Schwarzkopf, innocently surprised by all the fuss, pointed out that during his stay in Iran years before he had made many good friends, and now, in the course of a world-girdling holiday, he was here to see one or two of them. They included General Hassan Arfa, a staunchly patriotic Iranian but a bitter opponent of Mossadeq, who was suffering a 'diplomatic illness' on his estates at Larak, outside the capital, where he had a well-trained private army sworn to protect him; General Fazlollah Zahedi, another foe of Mossadeq, also in hiding on his estates; several officers in the police force; and of course the shah, Mohammad Reza Pahlevi. The US general did not seem to have any difficulty in obtaining speedy access to all of them.

As if realizing that Schwarzkopf's arrival was the catalyst of events, Mossadeq set out to quicken the pace of what was now evidently a complete takeover of the country. Cheered on by Tudeh mobs, he announced the abolition of the Majlis, the Iranian parliament, which was, the old man said, nothing but 'a nest of thieves'. He then declared that he would hold a referendum to give the people the opportunity of approving his action. He also asked for power to deal with 'sinister elements in the army, the police, and with other brainless agents of international reaction'. He made it quite clear that the latter phrase was a reference to the shah.

There were fights at the polls and the few who dared to vote against Mossadeq were badly mauled by the crowds. But in truth, the old man in the pyjama suit did not really need to manipulate the referendum votes, for the masses were with him, even if the army, the police and the landowners were not. But the strong-arm methods of the government guards gave the shah the opportunity of intervening.

Being a constitutional sovereign [wrote General Arfa later] the shah could not dumbly assist at such a tragic farce and he understood that this tragicomedy [of a referendum] was leading the country directly to chaos and anarchy, from which it could emerge only to fall into the hands of the Tudeh.*[2]

The shah therefore used his constitutional right and dismissed Mossadeq and announced the appointment as prime minister in his place of General Fazlollah Zahedi. No one (particularly the shah and General Schwarzkopf) expected Mossadeq to obey the order. As soon as he had made his announcement the shah took off from Teheran to join Queen Soraya at Ramsar, where he had a palace on the Caspian Sea. General Zahedi did not

* The Tudeh was the Communist party.

even bother to emerge from hiding or attempt to assume the post to which the shah had appointed him. He was quite right. The emissary who was sent by the shah to tell Mossadeq that he had been sacked was immediately flung into jail by the old man, who announced to the nation over the radio that he had refused to obey the royal order. A flood of anti-shah propaganda burst into print in the Communist newspapers and mobs swarmed into the streets execrating his name. At the suggestion of the US ambassador, Loy Henderson (now back in the country from his trip to Switzerland), the shah called for his private airplane, and, taking Queen Soraya with him, flew off into exile.

While the royal party was en route to Rome,* Mossadeq ordered a full-scale campaign to be unleashed against the monarchy. All the shah's pictures in government offices, cinemas and shops were removed, and the mention of the shah's name in morning and evening prayers in military units was forbidden. On 18 August 1953, the Tudeh party surged into the streets and starting razing to the ground all statues of the shah and of his late father.

It was the signal for which the anti-Mossadeq forces had been waiting. First, four hundred cadets at the Military Academy camp at Aqsassieh announced they were going on a hunger strike to protest against the insults to the shah. Then crowds of peasants, armed with staffs and daggers, picks and bicycle chains, suddenly emerged at the southern entrance to the city, and the fact that they came from the direction of General Arfa's estate was not without its significance. As they moved towards the centre of the city they gradually began to form into a procession, and were quietly joined by what General Arfa called 'many off-duty army and police NCOs'. Suddenly, thousands of anti-Mossadeq leaflets appeared. Army units and police squadrons ordered by Mossadeq to disperse the insurgent crowds joined them, and 'at ten o'clock, Parliament Square was full of people, and improvised orators standing on the base where the statue of Reza Shah had been, addressed the delirious crowds who shouted continuously: "Shah! Shah!". . . . At twelve o'clock most of the town was completely in the

* Where the Iranian ambassador refused to meet him – and lived to regret the lack of courtesy.

hands of the loyalists, the ministers having all disappeared, hidden nobody knew where.'[3]

Only around Mossadeq's house did an army unit resist the pro-monarchist mob, firing into them with tank guns and machine guns as they surged forward. Two hundred were killed and more than five hundred wounded. As the old man peeped out through the shutters at the screaming mob below him, he asked an aide: 'Where are all those who voted for me at the referendum?'

'*Bad bord*' (Gone with the wind), the aide replied.

By four o'clock it was all over, and General Zahedi, who had come out of hiding to lead the revolt, cabled the shah to return from exile. Mossadeq, still in pyjamas, was arrested in bed and taken to jail, to be subsequently tried for treason. The Soviet economic mission hurriedly returned to Russia by the next plane for Baku, and a ruthless purge of the Tudeh party began in all parts of the land, not least in the oilfields and Abadan, where thousands were imprisoned and hundreds subsequently shot.

SITUATION RESTORED. AS YOU WERE IN IRAN, announced the headlines in the British newspapers. In London the directors and staff of Anglo-Iranian got ready to take over once more, convinced that it would soon be business as usual.

But it turned out to be not quite that simple. For one thing, not even the anti-Mossadeq elements in Iran could stomach the idea of having foreigners back in control of their oil. For another, there was the question of the CIA. Thanks to General Schwarzkopf, the coup had been quite successful. But it had been extremely costly, and millions of dollars had been paid out in bribes. It hadn't all just been done for love of the Union Jack. What did the United States get back as compensation?

It was Herbert Hoover Junior who first thought up the idea of a consortium. He was President Eisenhower's adviser on petroleum affairs, and the moment the shah was safely back on the peacock throne he flew to Teheran to look over the oil situation. He had the field to himself for the moment. The Russians had taken temporary fright and withdrawn their economic and petroleum experts, though they would be back very shortly. The British did not even have diplomatic representation,* and the nearest Anglo-Iranian oilmen were

* After Mossadeq broke off relations with Britain in 1952, the US represented them in Iran.

waiting across the border in Iraq but were not allowed into Iran for the moment.

Hoover was well aware of Anglo-Iranian's attitude towards the concession now that Mossadeq was gone. It was status quo ante. As the company's historian put it later:

In London Lord Strathalmond and his colleagues from the day the company left Persia stuck their toes in with stubborn determination in defence of their rights, both legal and moral. They were determined that . . . the company's position should be acknowledged and respected.[4]

But Iranian pride was also involved. When Hoover reached Teheran he discovered that if Mossadeq had departed from the scene many of his ideas remained. The country was in a mess, deeply in debt, but in his talks with the shah and the prime minister, General Zahedi, Hoover discovered that both of them were determined on two things. Never would they allow Anglo-Iranian to come back as the sole controller of Iranian oil supplies. And never would any other foreigners be permitted to operate in the oilfields unless they first formally acknowledged that land and oil belonged to the Iranian people.

From Hoover's point of view, it was a situation pregnant with possibilities. What better method of bringing the two conflicting elements together than by using American petroleum interests as a bridge? The Americans would acknowledge Anglo-Iranian's prior claim to the concession, but would henceforward come in as the dominant partner, and perhaps invite others in to join the new line-up; the French, for instance, and the Dutch. An international consortium, in fact. Then they would go along to the shah and demonstrate to him that now no one company or nation would be exploiting Iran's oil resources, and that the new consortium, anyway, was prepared to acknowledge Iran's prior sovereignty over its petroleum.

To his amazement, not only were the British dead against the idea – they simply thought that America was trying to muscle in on a British oil monopoly, as indeed it was – but the American oil companies were not exactly keen on the idea, either. What Hoover had not realized was how well the big US oil combines had done out of the Iranian disaster. Aramco's oil production and sales from Saudi Arabia had gone up by fabulous amounts, and so had the figures in Kuwait and Qatar. The prospect of Iranian oil coming back on to the market, even though they would, as part of a consortium, make money out of it, filled the other American companies with more dismay than joy. In 1954, the economic climate was such that they were producing all the oil that the world could buy – and the return of Iranian oil to the market, as the experts saw it then, could only mean that production would have to be cut down in Saudi Arabia, Kuwait, Qatar and other Middle Eastern areas to accommodate it. What would the Ruler of Kuwait, the Emir of Qatar and the debt-ridden King Saud, all of them voraciously demanding more and more from their oil revenues, say when they were told that their incomes would henceforth have to be cut instead of increased?

In the event, it didn't work out like that at all. The world was soon lapping up all the oil it could get as greedily as the sheikhs were lapping up money, and not even the return of the Iranian fields would be able to satisfy all the demands. But even if the major oil companies' show of reluctance was not a deliberate ploy,* it certainly went a long way towards mollifying Anglo-Iranian over the coming loss of its monopoly in Iran. By the summer of 1954 the British had accepted the idea of a consortium, and would, in addition, be paid £34,500,000 negotiations with the Iranians. The line-up had been agreed. Anglo-Iranian would take 40 per cent of the shares in the new consortium,* and would, in addition, be paid £34,500,000 immediately by its partners as compensation plus 10 cents a barrel on all oil exports until a total sum of $510,000,000 had been reached.† The remaining 60 per cent would be divided up between the other members. They were not exactly newcomers to the Middle East oil game, or to each other. Royal Dutch-Shell would take 14 per cent. The Compagnie Française des Petroles got 6 per cent. And the remaining 40 per cent would be split (8 per cent each) among the US majors: Standard of New Jersey, Socony-Mobil, Standard of California, the Texas Company, and Gulf Oil Corporation.

'At least we won't be working with strangers,' said an Anglo-

* They had to be assured by the Justice Department they would not be contravening anti-trust legislation before they went in.

† The value of the company's Iranian concession was valued in 1953 – including the refinery at Abadan – at £550,000,000.

Iranian director after he had read through the list. If the Justice Department hadn't decreed otherwise, it would have looked uncannily like an international cartel.

A delegation set out for Iran pledged to get an agreement as urgently as possible and put the oilfields back into operation. But by the time they got there, the Iranians had recovered their nerve. And the Russians were back in the game.

With its economy in ruins and its anti-Western elements either in jail or dead, it might have been thought that Iran would make the quickest deal possible with the new consortium and start paying off its debts with oil revenues. But things had changed in Teheran, and so had the young Reza Shah. He had scurried out of the country in order to allow other people to overturn Mossadeq, but he returned not so sure that the old man hadn't been right after all. Too stubborn, obdurate and xenophobic, perhaps, but in principle a true Iranian in standing up for his people's rights. It was for this reason that he counselled the new government not to press too hard for condign punishment of the defeated ex-premier, even though he had conspired to topple the shah from his throne.* To his intimates, the monarch confessed his shame and humiliation that a change in government in his own country had had to be manipulated by the CIA, the intelligence arm of a foreign power. His courtiers noted a new firmness and certitude in his manner. He was no longer vacillating or fearful; he had even come to grips with the facts of his royal marriage, and he had already decided to give his people an heir even if it meant changing his queen for one more medically likely to give him the son he needed to ensure a successor to the throne.† As for the oil concession, he called in the premier, General Zahedi, and the finance minister, Dr Ali Amini, and told them: 'The Americans are coming in search of an easy bargain. Anyone who gives it to them will answer to me – and to the Iranian people. I want it known as widely as possible by my people that the oil will stay Iranian, and that they will decide the

---

* Whereas others were savagely treated, Mossadeq was given a sentence of three years' imprisonment and received special treatment.

† Queen Soraya, who failed to bear him a child, was subsequently divorced by the shah and sent into exile. He then married a young Iranian student (Soraya was half-German, half-Iranian) who has since borne him a crown prince.

punishment of any negotiator who fails to bear that in mind'.[5]

He then stole a leaf out of Mossadeq's book, and instructed the prime minister to let it quietly be known to the Russians that they were welcome to send their petroleum experts back into the country, and that they need not entirely lose hope yet of securing a concession. He presumed, rightly, that the Americans would hear about it.

That royal audience was given in February 1954, and from that moment on the stage was set for horse-trading of a kind that none of the Anglo-American delegation had bargained for. Four weeks of preliminary negotiations, led by Ronald Hardman, vice-president of Standard Oil of New Jersey, ended with the Iranian delegation handing a memorandum to the Anglo-American delegates just before they flew back to London to report. This, the so-called Airport Memorandum, contained demands so stern and uncompromising that, to quote a Standard of New Jersey official, 'practically everybody in London said: "It's finished. We'd better forget about Iranian oil. You'll never make an agreement with those guys." [6]

But if the oilmen were ready to give up, the British government was not, nor, at its urging, was the US government. The British needed the oil revenues and the compensation from their new-found partners. The US government wanted to make sure that Iranian oil was kept out of Russian hands.

'They put the heat on us,' the official said. 'They insisted that we go back and have another try.'

There was only one man among the Americans who believed that an agreement was possible. His name was Howard W. Page and he was representative in London of Standard Oil of New Jersey. Son of an oilman, born and brought up in the California oilfields, trained as a petroleum chemist, the veteran of several tricky and complicated concession deals,* Page walked into a conference room with the friendly air of an honest man (which he was) prepared to do a fair deal with men of equal probity (which his opponents often found disarming). When the Iraq-IPC Agreement of 1952 was in its final draft a meeting took place in London of oilmen and their British legal advisers who went through the agreement paragraph by paragraph, after which each was asked for his comment. When

---

* He wrote the terms of the 1952 agreement between the Iraq government and the Iraq Petroleum Company.

Howard Page was asked for his opinion, he said, to the astonishment of them all: 'There's only one thing wrong with this agreement. It won't be accepted by the Iraqi government.'

Since it contained everything which the Iraqis had insisted on, the lawyers indignantly asked why not.

'Well, I'll tell you,' said Page, calmly. 'I'm neither a lawyer nor am I British, and that makes it difficult in one way and easy in another. This sounds to me just like the contract I had to sign when I rented my apartment in London. Now that starts out assuming I'm a crook and I have assured them in print that I am not a crook. Now I didn't like that one bit, and if I didn't have to have a place to live in, I'd have told them to go and take a running jump at themselves. Now this agreement here is the same.'

He paused while they asked for specific objections, and then went on: 'Let's just take this paragraph. It says that the Iraq government shall not under any circumstances do this. In the next paragraph, it says that the Iraq government shall not under any circumstances do that. Now that's no way to write a contract with a friendly government. You can get the meaning in without throwing it right at them that you think they can't be trusted.'

There was a silence, and then the chief of IPC's legal advisers said: 'If that is the way he feels about it, perhaps it would be better if Mr Page were to rewrite the contract.'

'Dammit, I will,' said Page.

He went back to his apartment and sat up until three a.m. the next day restructuring it paragraph by paragraph.

'Basically,' he said later, 'I didn't change anything. I mean I didn't alter the sense one bit. But when I handed it over next day, and they went through it, the chief legal eagle said: "Yes, this does sound better. It's also rather more legal." '[7]

But more important, the Iraq government liked it – and signed.

Howard Page had spent six months studying the terms of the uncompromising Airport Memorandum from the Iranian government, and at the end of that time he was convinced that an agreement was possible, no matter how unlikely it looked at first or second glance.

'I thought there were just enough elements on both sides,

just enough flexibility, to make it feasible,' he said. 'I thought that we could lose some things we could afford without giving away what we couldn't afford. Also my experience in Iraq had taught me that if you want an agreement you can get it. I sensed that the Iranians wanted and needed an agreement, but that their representatives were asking for the sky because they were scared of not getting the sky – of what would then happen to them at the hands of the Iranian mob. Then I heard from Dennis Wright, who was the new British chargé d'affaires in Teheran, an absolutely excellent diplomat, that he believed there was more flexibility in the Iranian position than appeared on paper. It was at this point, because I knew the Airport Memorandum like my own face, and because I was an American and the Iranians wouldn't deal with a Briton, I was selected to head the delegation that would go back to Teheran and reopen the talks.'

The delegation flew out in chartered planes in April 1954. Page took William Snow of BP* and John Loudon of Royal Dutch-Shell along with him, and both were senior oilmen, but as an American and as an experienced negotiator there was never any question of anyone but Page being in charge. The Iranians had put Dr Amini at the head of their negotiators, and it was a wise choice. 'I would say that there were not more than ten men in the world who knew more about oil than he did,' said Page later. Amini was a brilliant scholar, he never lost his temper, and he had a marvellous sense of humour. But his greatest advantage was his courage. Unlike his subordinates, fearful of royal wrath or mob violence if they showed any weakness, he was prepared to concede points as well as demand them.

'He was the one man,' said Page,' who would take responsibility. He would keep his eye all the time on the object of getting an agreement, and he would be prepared to say, well, we can give a little here, providing you give a little there. I'm sure that without him we would never have made it. None of their experts had enough guts – and you can't blame them. There was always the fear of assassination. Feeling was running high, and both delegations had bodyguards.'

A few hours after Page and his delegation arrived in Teheran

* Since it no longer had sole possession of the Iranian concession, Anglo-Iranian had now changed its name to British Petroleum, by which it has been known ever since.

G

he was called to the US embassy for a conference with Herbert Hoover Junior. Hoover informed him that the Russians were back in the game, and were desperately anxious to know what the Anglo-American delegation was going to offer the Iranians, and what tactics they would adopt once the talks got under way.

'Wherever you talk,' Hoover said, 'the Russians are going to bug you. Now I recommend you to come here and have your conferences in the embassy. We've debugged this place, but we can't debug other places.'

Page declined. He didn't particularly want the US government to know what the delegation was planning, either. He knew his British and Dutch colleagues would be even more embarrassed than he was at talking under the protection of a diplomatic mission. But during his conversation with Hoover, an embassy electronics expert came in, and he made one remark which Page remembered.

He said: 'You know, the only thing that can stop these guys from hearing what you're saying is running water.'

Page knew just the place. The negotiators were living at the Darvan Hotel at Shimran, in the foothills overlooking Teheran, and what the American had noticed in his strolls was a garden.

'There was a house with the garden, but the house wasn't worth a damn,' he said. 'But the garden was very nice. It had a swimming pool and running water – and where the water went out of the swimming pool and into another pool was just the place for our conferences.'

The house was rented, a gardener's cottage was fixed up for the secretaries, and they were in business.

The negotiations which everyone (particularly the British) thought were going to be so easy – for after all, the Iranians had been forced to the conference table – were both complicated and protracted. The Iranians had a poor hand of cards but Amini was a master player. And it turned out, though the delegates didn't discover it until later, that the Russians had slipped an ace into the pack. One of the Iranian team, also a member of the Majlis and one of the committee which would, once an agreement was reached, have to sign the documents, was under Russian control. Mr Salim* objected to any sign of flexibility on Amini's part, he talked endlessly for hours on end

* This is not his real name.

about Iran's 'sacred oil' and her 'inalienable rights to her precious heritage' and he procrastinated whenever possible, driving everyone to distraction and prolonging the sessions for days and weeks.

The talks which had begun in April dragged on through June, July and August. The shah began to be impatient. He called Page and Amini to an audience at the palace and told them that he was summoning the Majlis for a week hence, at which time an agreement must be ready for the members to examine and approve. Either all details would have been ironed out by that time, and all objections resolved, or he was cancelling the negotiations and would take 'other steps'.

'You have seven days, gentlemen,' he said.

In the next seven days, Mr Salim surpassed himself. By this time a draft agreement had been painstakingly drawn up in English and an Iranian translation was being made, and since both would be valid in law it was necessary that an exact translation should be drawn up. This was difficult because there were no Persian words for certain technical points and processes. Furthermore, Mr Salim took it upon himself to supervise the translation, and he managed to argue over every phrase at the rate of about one an hour.

Page knew that the shah meant what he said about the time limit.

'The Persians always have this thing about time, and they are unforgiving if foreigners don't keep to it,' he said.

It would take three days to get the Persian-language version to London, the Hague and New York, where the heads of the companies concerned would sign it, and then back to Teheran. Page had a charter plane standing by, but thanks to Salim's procrastinations, the pilot ran out of rest-time. The charter plane had to be scrubbed. Then someone mentioned that the regular KLM service was due to leave in forty-eight hours for Amsterdam, and it so happened that the Dutch ambassador would be aboard. Would he take the precious agreement with him to Amsterdam, where Royal Dutch-Shell representatives would be waiting and would rush it to the others? He would take it? Good.

Despite last attempts at delay from the indefatigable Mr Salim, the agreement was finished at last. The Majlis was informed and the members began to assemble, waiting to hear

the agreement read to them the moment the Iranian committee had signed it. The committee was summoned to an anteroom in the parliament building, where everything was ready
for signature. Finally they were all there, including Mr Salim,
making no secret of his resentment. But Page wasn't worrying
about him any more.

'I had got this thing nicely wound up,' he said, 'and we had a
good hour and a half to spare before we took the contract down
to the airport. One by one, the committee members sat down
and began to sign. And then, just before it came to his turn, Mr
Salim excused himself and said he had to go to the men's room.
And he didn't come back. He didn't come back for an hour and
fifteen minutes. We were going crazy – and nobody did anything about it. We just hung around and waited, and all I could
think about was that plane on the airport, waiting to take
off. If we missed it we might as well throw the contract in the
ashcan and jump in with it.'

Eventually, Mr Salim returned. He made no apology or
explanation. He simply sat down and signed his name to the
documents.

'We were so late now,' said Page, 'that I didn't even stop to
wrap the papers up. I just grabbed them, and I rushed out without saying goodbye to anyone. Luckily I had some paper and
string in the car, and I told the driver to get going. He was one
of the wildest drivers in Teheran, the roads were pretty rough,
and I was bouncing around trying to tie up the documents.
When we got to the airport a KLM man was waiting for us. He
jumped into the car and said: "Drive out on to the runway." He
wigwammed to the control tower and we sped away. The plane
was on the runway, already warmed up and just ready to take
off. My driver was now so excited that he surpassed himself,
and we went along the runway like cats with cans on their
tails. It was one of the large old piston planes, and I don't think
the driver saw those propellers going round at all. We were
coming head on to the plane and just about to go under the
wings, when I leaned over and wrenched at the steering wheel,
just in time to swerve out of the way of the whirring propellers.
We screamed to a stop and I flung myself out, package in hand.
The door of the plane opened, I tossed the clumsily tied
bundle in, and as the plane passed us I got a thumbs-up sign

from the ambassador at one of the windows. That's how we got the contract through.'

The agreement between the consortium and the Iranian government was approved by the Majlis in August 1954, and formally ratified in October of the same year. British Petroleum had no reason to complain about the settlement, in the circumstances. Its partners were paying the company by instalments £200,000,000 in compensation for its loss of the monopoly. The Iranian government agreed to hand over £2,500,000 a year for ten years to BP, also in compensation. As for the partners, they were now operators of one of the biggest oilfields in the world and they had got it at a bargain price, which they could easily pay out of income.

For the moment, the Iranians were not grumbling, either, though they would do so later. In return for their compensation payment to BP, they were given the ownership of all the company's physical installations in Iran for producing and refining oil, and they would also become proprietors of any other installations which the consortium built in the future. These would henceforth be the property of the National Iranian Oil Company, the nationalized body which had been created by Mossadeq after he took over Anglo-Iranian.

The consortium (through two newly-registered companies) would get the oil out of the ground and then sell it to the NIOC, which would in turn sell it to the separate trading companies set up by each of the partners. And from the profits which these trading companies made in marketing the oil abroad, the Iranian government would take 50 per cent in taxes.

It was a profitable arrangement for everyone. Too profitable for the companies, the Iranians would say later. But for the moment everyone was pleased. Iran's oil was owned by an Iranian national company. There were Iranian members on the boards of all the subsidiary companies of the consortium. And in the next three years, the Iranian government received $490,000,000 in direct oil revenues alone.

There was one man who expected to benefit from the establishment of the Iranian consortium, and that was Enrico Mattei of Italy. Unlike France, Italy was one of the major European countries with no oil production of its own, and Mattei, as head of the nation's chief petroleum company, AGIP, had channelled all Italian oil purchases into Anglo-Iranian. When Mossadeq nationalized the company and sought world outlets for the fuel he now had on his hands, he had made a direct approach to Mattei. But, as his biographer and former business associate, P. H. Frankel, wrote afterwards:

It is worth recalling that during the years of conflict of Anglo-Iranian with Mossadeq, Mattei proved to be entirely loyal to the former, AGIP's exclusive supplier; even when the Italian government acquired 'stolen oil' from Dr Mossadeq through a somewhat obscure firm by the name of SUPOR, Mattei would have nothing to do with them. Thinking that such an attitude had earned him gratitude which would be adequately expressed on a suitable occasion . . . Mattei appears to have inquired whether he would be included in the consortium to be formed.[8]

He was kept waiting for an answer until the consortium was ready to do business with Teheran, and then Mattei was given a flat no by the participants.

It was apparently said that only companies which were already holders of concessions in the Middle East would be eligible and that to open the door to any others would mean creating a precedent which would call forth such a spate of claims of would-be participants as to render any solution impracticable.[9]

Mattei was afterwards to maintain that he was cheated by the big oil companies, and that they had made promises to him which they afterwards did not keep. As to that, Howard Page says:

'At no time did Mattei ever approach me, nor did I ever hear that he had approached others, to get in on it [the consortium]. I would have known if there had been a confidential approach. Certainly if Jersey had been approached I would have heard of it. So how he had the nerve to say what he did about us I don't know. There was no need to pay him off for anything.'[10]

Mattei might have taken his rejection with better grace – putting it down to the determination of the 'big boys' to keep Iran to themselves – if, a few months after the formation of the consortium, there had not been a new development.

When the idea of American involvement in the Iranian consortium had first been mooted, the companies had approached the US Justice Department to find out whether their participation would be a breach of the anti-trust laws. They were assured that they would not, in the circumstances, be prosecuted for transgressing the law. But after the contracts were signed, the Justice Department changed its mind. Responding to pressure from the independent oil companies' lobby in Washington, it abruptly announced that legal approval of American participation would be withheld unless independent US oil companies were allowed a share of the partnership. The five US majors involved in the consortium thereupon met and decided to avoid trouble by ceding five per cent (one per cent each) of their share of the consortium. Nine separate independent US oil firms, seven of which had never been involved in the Middle East before, found themselves participants in the Iranian bonanza, and with no entrance fee either.* As P. H. Frankel remarks:

It was almost grotesque to see that for no reason other than the desire to pay lip service to free competition, a few American investors were, by courtesy of Washington, given what amounted almost to a free ride, although the oil they were to get played no functional role in their activities, whereas a country like Italy and a refiner/ distributor like AGIP were to remain outside.[11]

For the newcomers it was 'a licence to print money,' as Mattei put it. He never forgave the consortium companies for the snub. From then on until his violent and mysterious death in 1962, he embarked on what Frankel describes as 'a collision course' against the major petroleum combines.

Meantime, the nine American newcomers to the consortium sat back (they were not expected to take any part in the operations) to enjoy their unexpected profits. It was just like Gulbenkian all over again – except that, unlike the Armenian entrepreneur, the lucky nine had done absolutely nothing to earn their five per cent.

---

* Of the nine companies, two were Getty operations: Getty Oil Company and Tidewater Oil Company. The others were: Richfield Oil Corporation, Standard of Ohio, Signal Oil and Gas Company, Atlantic Refining, Hancock Oil Company, Dan Jacinto Petroleum Corporation, and American Independent.

PART FIVE

The Breakthrough

17 · Bird Sanctuary

In 1953 Captain Jacques Cousteau, the French undersea explorer, sailed his research ship, *Calypso*, into the Persian Gulf and anchored off a small island about eighty miles from the shores of Abu Dhabi. The island was called Das, and it was at that time one of the wonders of ornithology. Each year millions of terns swarmed towards it from all parts of Asia and Africa and laid their eggs in clutches along the foreshore. Cousteau and his crew smelled the reek of guano from fifteen miles away, and were thrilled by the aerobatics of these myriads of sea-swallows as they soared into the brilliant blue sky and then dived at the shoals of fish boiling close to the surface of the Gulf. For the next few months Cousteau and his divers were overside, mapping and photographing the sea bed below them, and soon they were joined by a vessel of the Geomarine Service International and a Swedish submarine charting ship.

For Cousteau it was one of the most fascinating missions of his distinguished and exciting career, for he found the area a paradise of animal, bird and fish life. The shores of Abu Dhabi and the islands were rich in foxes, hares, gazelle and spectacular giant lizards whose dayglo colours changed with chameleon swiftness each time they crouched beneath a patch of scrub or rock. The seas were alive with shark, squid, prawns, turtles, sailfish, horse mackerel, great schools of barracuda and massive shoals of multicoloured fish of all kinds. The air quivered with the sound of wings.

Yet Cousteau was downhearted because he knew that the success of the mission on which he was engaged would doom this tract of pulsating activity to a slow death by oil. This was the decade of expansion by the oil companies into the shallow waters of the Persian Gulf, and the five hundred miles of sea from the Shatt-al-Arab in the north to the Strait of Hormuz in the south-east had been divided up between the states whose foreshores were lapped by the waters. In Abu Dhabi the main concession had been granted to a company called Abu Dhabi Marine Areas (ADMA) in which British Petroleum and the Compagnie Française des Petroles were partners, and already an advance headquarters had been established on Das. The first clutches of precious eggs had been trodden underfoot and clouds of furious terns beaten off by labourers. By the end of the year, a 4,000-ton drilling barge named the *Adma Enterprise* anchored off the island and lowered the four retractable legs with which it was equipped down to the sea bed at a spot Cousteau had designated eight fathoms below. The *Adma Enterprise*, built in Germany, was a remarkable vessel capable of submarine drilling down to a depth of 15,000 feet, but it did not need to go that deep to strike oil. Biting through the sand and rock, it hit a pool at 5,500 feet, and oil began seeping to the surface of the Gulf, and the first dead fish floated on the water.

By the time Cousteau sailed away in *Calypso*, the bulldozers were biting into the small jabals of Das Island and carving out space for hutments, tank-farms, a landing ground and a small golf course. Crude oil was rolled into the ground to lay the dust, and the smell of petroleum took the place of the reek of guano. Tern eggs were picked and eaten as delicacies by the newcomers, and the birds that stayed obstinately on the island became targets for oil company marksmen. Soon even the shape of the island would change as the hillocks were shaved down for building sites, and every trace of vegetation would disappear, smothered by oil and the paraphernalia of a petroleum collecting centre and loading point. Where swordfish once broke the surface, gas fires would now flare across the Gulf, and one of the most spectacular bird islands the world had ever known would be no more, killed by the civilized world's greed for petroleum.

The Trucial States of Arabia in the 1950s were ruled by an oligarchy of rapacious sheikhs whose power of life and death over their subjects was – since the maharajahs of India had lost their power – unique in the world. Rarely has there been a greedier – or more shortsighted – group of men.

Abu Dhabi was on the brink of one of the most fabulous breakthroughs in the history of oil exploitation, and already large revenues were beginning to flow into the sheikhly coffers. Yet Shakhbut bin Sultan, the ruler, had no intention of using his growing wealth to improve the lot of his people. He preferred to keep it in gold bars under his bed and in stacks of notes in an anteroom off his living quarters.* He was in truth a mean and miserable old man, not given to cleanliness (the virgin slave girls for whom he had a passion must have found him a loathsome burden), suspicious, jealous, and arrogant, subject to black rages, when his voice rose and he shrilled like a child. Anyone who tried to get his signature on to a document learned just how cautious he could be. His British adviser recalls an occasion when one of the managers of a large American bank was attempting to gain an entry for his branch into Abu Dhabi:

> The bank manager was accompanied as interpreter by a Lebanese of distinction, who was on his knees before the ruler praising him, his wisdom, generosity and charm. It would really have been difficult for the ruler not to accede, and it was only when the bank manager put his hand to his inside pocket and started to extract a paper which was obviously intended for the ruler to sign that Shakhbut drew back with the look of a trapped animal. The Lebanese was forward on his knees like a praying mantis, hands raised in supplication. The ruler sat terrified on the back of the sofa and I shook my head to the bank manager signalling him to put the paper away quickly. So keen was he on his object that he was a bit tense and tried to press on, resulting of course in the ruler's even greater apprehensions and more solid resolution not to sign the paper.[1]

Shakhbut even in the 1950s was receiving something like £20,000,000 a year in oil revenue, and with a tenth of that sum he could have brought some measure of medical aid and education to the 15,000 men, women and children who made up the population of Abu Dhabi. But that would have meant parting with his precious notes and bars of gold, and the sheikh could not bear it.

The Trucial States of the Persian Gulf† were all under the paternal control of the British. British influence had been

* When he was ousted in the 1960s it was discovered that several square yards of money, worth at least $2,000,000, had been eaten by rats.

† In addition to Abu Dhabi, there were five other trucial states: Dubai, Ras el Khamma, Um al Kawain, Sharjah, and Trucial Oman.

dominant in this area for a century, ever since, in 1853, British warships had attacked the buccaneers who terrorized these waters, and forced the rulers of the states along the Gulf shore to sign treaties pledging themselves to cease their attacks upon passing shipping. The name of the area changed from the Pirate Coast to the Trucial States, and there had been British advisers on hand from that date onwards. They directed the foreign policies of the tiny sheikhdoms and provided officers and NCOs to stiffen their small defence forces. They pledged themselves not to interfere in the domestic affairs of their desert brood, and by so doing condoned rulers who at their best were petty autocrats and at their worst bloodthirsty tyrants.

Oil and the revenue it produced was beginning to bring out-siders and outside ideas into these stuffy little kingdoms, but rare were the sheikhs who welcomed them. Sometimes a British political agent longed to see the money from oil revenues used for worthwhile projects, and in Abu Dhabi the local British agent, Hugh Boustead, pressed Shakhbut to employ part of his wealth in changing the country's most fertile area, Buraimi, from a disease-ridden, one-crop oasis into a market-garden showplace of Arabia. In furtherance of his hopes, he produced an agricultural expert and a detailed scheme for the sheikh's inspection. He might have been talking to a deaf man, for Shakhbut would hear nothing of his plans.

'My people know more about agriculture in their own land than you can tell them,' he said to the crop expert. 'The budget [for the agricultural scheme] is a pure waste of money. We do not want it and we will not have it. What can you tell us that we do not know?'[2]

It so happened that when the interview took place the ruler's brother, Sheikh Zaid, whose home was in Buraimi, was present, and Boustead caught the look of anger and frustration in the younger Arab's eyes.

Shaikh Zaid was very disappointed [he wrote later]. He took on a very able Pakistani agricultural officer, Mr A. H. Khan, and paid him from his own funds, but there was no supporting budget, and as a result agricultural development has been held up completely . . . The most disheartening feature of all was [the ruler's] simple refusal to give his own people, a pastoral, agricultural and fishing society, the easy assistance which his growing wealth could provide.[3]

One day there would be a confrontation between the ruler and
Sheikh Zaid, as there always had been between brother and
brother in the bloody history of the ruling families of the
Trucial States. No one could guess at that time who would win,
but Boustead had no doubt in his mind which of the two sheikhs
the people of Abu Dhabi would prefer as their ruler.

If they had to have a ruler, that is.

Oil. The greedy sheikhs dreamed of it, and of the wealth and
power that its discovery would bring them, and no one more
so than the bearded tyrannous ruler of the wild kingdom of
Muscat and Oman, whose shores stretched all the way from
the Gulf of Oman and the Arabian Sea to the Indian Ocean.
Sultan Said bin Taimur was the kind of Arabian sheikh that the
playwright must been thinking about when he wrote *The
Desert Song*. In his Arab clothes and with his flowing beard he
looked somewhat like a Biblical prophet, but he was also wont
to lave himself with unguents, groom himself meticulously, and
dress in Western clothes cut by Savile Row tailors. He spoke
flawless English which he had learned at school and university
in the United Kingdom, and he had a devastating charm which
he used – often to a degree they lived to regret – on young
European women. He was also a despot who imposed upon his
people a regime so rigidly theocratic that even Saudi Arabia
in the heyday of the ulemas (the fanatic Muslim priests) had
known nothing as pressingly sanctimonious. In his hot foetid
capital of Muscat no music was allowed and no dancing.
Women were permitted on the streets only after darkness, and
then only in the company of bodyguards. There was a curfew
by night when the gates of the city were shut tight, and before
that all who moved abroad must carry a lantern with them
(and by lantern the sultan did not mean an electric torch).
Wrongdoers were mercilessly lashed or bastinadoed or slowly
tortured to death.

But the sultan's greatest crime against his people was his neg-
lect of them. In no other country in the Middle East were
curable diseases so rampant, syphilis, dysentery, tuberculosis,
malaria, fearsome intestinal worms, all of which tribal quacks
treated with hot irons, and then rubbed the burns with camel
dung and exposed them to the flies. There was but one hospital
run by a few brave American missionaries, allowed by the

ruler because they cost him nothing. It was a land of prohibitions, ferocious prejudices, fanatical cruelty; and the worst sufferers, their miserable lot approved by the tyrant ruler, were the women. An American missionary said:

Once again comes the agonizing cry : 'Come quickly, doctor, my wife is in labour and cannot deliver.' Is this her first baby? No, her second. She had no trouble the first time. Atresia : that is what the doctor suspected and found. Atresia, a condition which causes the death of many a mother and her unborn child . . . is caused by fear—fear of divorce. Atresia is a condition where the normal tissue of the vagina has become inelastic scar tissue. When this condition exists the child cannot be born. The woman in labour suffers agony for in her body the irresistible force of birth contractions propels the foetus against an immovable barrier of rigid scar tissue, which has caused the birth canal to tighten like a purse string. Suddenly everything rips open.

How was this scar tissue formed? The answer is that it results from the packing of rock salt into the vagina after childbirth—after the birth of the first baby. Why salt? The answer—to contract the vagina lest the husband, not deriving the satisfaction from his wife he experienced before delivery, should divorce her . . . Surgery alone can save the mother and child. Every year a score of such operations are performed in the American Mission Hospital in Muscat.[4]

Only education could bring this barbarism to an end, and only education could persuade the sultan's subjects to seek medical care instead of savage tribal remedies for the loathsome diseases to which they were heir. But Taimur bin Sultan would have none of it. There was only one primary school in Muscat and that had few Arab pupils, but was, instead, largely used by the sons of Indian merchants. When the British put up a programme for the building of schools, he would have none of it.

'That is why you lost India,' he said, 'because you educated the people.'

But oil he wanted. He had travelled north to Kuwait and seen there the fabulous results, the gaudy opulence, which had been achieved through the rich revenue from oil. He had heard the clink of gold coins in the coffers of his miserly neighbour, Sheikh Shakhbut. He was convinced that in his vast kingdom of desert and mountain there must be equally rewarding troves of petroleum. But who would find them for him?

So far he had had no luck. The Iraq Petroleum Company was

still searching along the coastal plain of western Hajar, in Oman proper and though they would make valuable strikes in the years to come, so far their labours had produced no viable fields of oil. Further south, in the mountainous dependency of Dhofar, IPC had given up its concession at the end of 1950, defeated by the terrain, the absence of roads, the lack of water, and harassment from rebel tribes infiltrating from Saudi Arabia and the Yemen.

'Not too encouraging, is it?' said the sultan, one day in 1953.

He was talking in his seaside palace in Salala, in Dhofar, to a young American named Wendell Phillips. A year previously Phillips had led an archaeological expedition into the Yemen, which lies to the west of Dhofar, in search of the sites of ancient Sabean cities. Suddenly the party was attacked by rebel tribes and was forced to flee into Dhofar, leaving most of their equipment behind them. The American, called to an audience with the sultan, had recounted the story of his misfortunes and found the ruler, to his surprise, to be a sympathetic listener. In fact the two men established an immediate rapport. Phillips is a tall, pale, somewhat theatrical figure who likes to sport Arab dress and carry a couple of six-shooters in his belt, and he talks with the fluency and fervour of a revivalist preacher. He spoke of his explorations with such enthusiasm that Said bin Taimur was moved to unexpected generosity, and he invited Phillips and his party to continue their researches in Dhofar, and advanced him money to pay the cost of new equipment and native labour at the digging sites.

Now Phillips had come to tell him that the advances were used up and the expedition was on its way back to New York. Obviously taken aback by the imminent departure of a young man of whom he had become extremely fond, the sultan thought for a moment, and then suddenly said: 'We need oil in Dhofar. And by the will of God we shall have oil, for I am granting you the oil concession for Dhofar.'

Recalling that moment later on, Phillips said: 'I didn't know what to say at first. All I could think of to ask was how big was the concession, and the sultan waved his hand and said: "All of Dhofar. It is as big as your state of Ohio." It gradually dawned on me that Said bin Taimur was offering the rights of this huge piece of territory to me. I thanked him for his gracious gesture, but then, my wits coming back, I pointed out that

I knew nothing about oil, or oil concessions, or Arab oil concessions.'[5]

'You'll learn,' the sultan said. 'Now shall we begin work on the oil agreement, or would you prefer tomorrow morning?'

As Phillips wrote later:

When you are a penniless explorer and have no knowledge of the oil business, the possession of an oil concession presents certain problems. Where does one go from here?[6]

He flew to New York convinced that he had a surefire fortune in his briefcase, and that all he needed to do was knock on the door of the nearest oil company and offer it to them. But it wasn't quite as easy as that.

'For one thing, IPC had given up the concession after fourteen years,' Phillips said. 'True, they had never drilled a well in the whole of that time, merely made geophysical surveys, but their failure to do so was a strike against the terrain. Then Dhofar sounded so remote. I had to admit that when summer came it got really hot – though not as sticky as Muscat, where the climate is hell on earth – and you wallowed in floods of mud when the tail end of the monsoon hit. I soon learned that the cost of development would be enormous. Since there were no ports or docks, everything would have to be flown in and brought ashore by dhow. Time and again, oil company executives told me to hang the concession on the wall as an interesting decoration.'[7]

Then Sam Pryor, the Republican leader and a vice-president of Pan-American, joined forces with the young explorer and formed a company, Philpryor Corporation, to exploit the concession. Pryor had friends and know-how. In short order, he had assigned the concession to a holding company, City Services Corporation, and through them to another subsidiary company. Through the sale of units, Phillips suddenly found himself in possession of his first million dollars.

'I remember I had an account at the Chase Bank,' he said. 'A small account – it couldn't have been more than thirty dollars. So I went in and handed over the cheque I had just received from the brokers for a million dollars. Next thing I knew two policemen had grabbed me by the arms, because the teller had rung the bell. Well, I managed to persuade them that the cheque was genuine, and the next day the manager invited me to lunch, and he brought his daughter along. It was my first indication of what it was like to have a million – and of the potentialities of the oil business. And we hadn't even started looking for oil in Dhofar yet.'[8]

In 1957, at Marmul, thirty-seven miles inland from the Dhofar coast, the Phillips concession struck oil at 320 feet. It looked as if the company, after months of heart-breaking failure, had at last found a textbook structure and an oil reservoir at least eleven miles long by five miles wide.

When they broke the news to Sultan Said bin Taimur, he was overjoyed. He ordered an immediate reduction in the import duty on coffee, and allowed the curfew in Muscat to be put back by one hour. To his American friend, he said: 'Wendell, now you will be rich. How does it feel?'

'I am a modest man,' said Wendell Phillips. 'I shall not need to learn how to be a modest billionaire.'

But both of them would have to learn that in the oil business bonanzas are not always what they seem.

Kuwait was now the dream city that every Arab envied. In 1952 its population was 310,000 and its income from oil was $60,000,000. Two years later it was $210,000,000 and by 1955 it had reached $300,000,000. And each year more and more money flowed into the tiny state's coffers, until they were soon literally choking with it.

'At first we used to keep the revenue in an enormous safe in my office,' said Colin McGregor, who is now financial adviser to the Kuwait government. 'We stuffed in stack upon stack of overprinted Indian rupees, gold bars and sacks of sovereigns. There were millions and millions of them. The ruler's relatives would come in and pick up a million rupees at a time, and would get very snooty indeed when I insisted on a receipt. I remember once one of Sheikh Abdullah's most spendthrift brothers sent a message over saying he needed a huge sum of money in gold sovereigns and Maria Theresa dollars, and I put them in two large sacks and sent them back to him, but with strict instructions to the guards not to come back without a receipt. They came back an hour later with an old piece of paper on which was written: "Received: 2 sacks of money." '[9]

Sheikh Abdullah, the ruler,* was a shrewd, or rather a

* Sheikh Abdullah al Salim al Sabah succeeded his father in 1950.

200

cunning, man. He decided that if there was going to be such a flood of money flowing into his possession, a large portion of it should be channelled off into the pockets of his people, so that none could accuse him – as they did in the case of King Saud of Arabia – of keeping it all to himself and his cronies. Colin McGregor and other financial advisers persuaded him to keep one third of the yearly income as capital for investment, but the rest was lavished on the natives of Kuwait in proportion to their status. The ruler decreed that the squalid old town and port from which the pearl-divers had sailed, and the bazaars where their finds had been bought and sold, should be razed to the ground and a new city constructed to take its place. He sanctioned an immediate expropriation of land, but at a hand-some amount of compensation. Soon there were few Kuwaitis who did not benefit from the boom, which took the price of land in the 1950s to $200 a square foot.

Kuwait was henceforward to be a welfare state in which no citizen would lack for space and recreational centres while he was well or medicine and hospital care when he was sick. A large home and hospital centre was even constructed for old people – this in an Arabia where old folk are left to wither away and die in pain and squalor. Schools began to rise out of the sand, wide roads, handsome squares and traffic roundabouts; and soon cars flooded in to fill them with traffic. From now on, if you saw a man with a camel or an ass, he was not a Kuwaiti.

All this cost fabulous amounts of money (about £100,000,000 in the space of five years), and since few Kuwaitis knew anything about construction the city became the inevitable victim of foreign jerry-builders. Roofs fell in and roads buckled and split. Cheap materials collapsed under the onslaught of sandstorms or winter rains. Entrepreneurs upped their costs in the course of construction, rightly calculating that the Kuwaitis would pay and pay, and pay again, to get a new swimming pool or college or secondary school finished. By the end of 1956, the head of Kuwait's development board, who happened to be the ruler's brother, had to go to him and confess that he was £70,000,000 in debt on his budget. He was chided to go away and not spend so lavishly in future, but it was not until the 1960s that experts and accountants were called in to get some reality into the expenditure.

From Sheikh Abdullah's point of view, however, it was all in the good cause of keeping his subjects happy. As for his close relatives, he was prepared to pay them incomes of anything up to $1,000,000 a year, on the theory that a young man with money to spend – and the relatives were encouraged to spend – would not be likely to waste time on plotting assassinations (hitherto a family preoccupation) or coups d'état. To show his gratitude, one of the relatives organized a feast at which a feature was a bonfire started off by igniting a stack of hundreds of 5 dinar notes (worth £5 apiece), and which ended with a presentation to the ruler of a model of an oil derrick fashioned in solid 24-carat gold. In the meantime, guests consumed young roasted camels, each camel containing inside it a roasted sheep, and each sheep stuffed with roasted chickens, ducks and pigeons. It was afterwards said that each of the birds contained a boiled egg.

Along the shores of the Gulf can be seen today the traces of what the ruler's relatives did with some of their enormous incomes. They built palaces, with vast filtered pools, air conditioning, great marble chambers, filled them with expensive modern junk from European department stores, and then let them rot under the onslaught of sandstorms, humidity, ants, mice and rats. Private golf courses were laid out in the desert and planted with shrubs, imported at heady expense from Europe and the United States, and then allowed to wither and die for lack of water. The cabarets of Beirut, Cairo and Istanbul were plucked of their prettiest *poules* and brought to Kuwait to adorn the tawdry palaces, and for a time the streets of the city were suddenly alive with female voices chattering German, Hungarian, French and Rumanian from behind gossamer veils; and then a sudden rash of venereal disease started a wave of revulsion, and the foreign tarts were expelled. Back they went to the Levant, well content with the fruits of their labours in this hot land, loaded with gold bangles and precious jewellery.

The next wave of foreign female imports were rather more expensive, principally because their backgrounds were more respectable. There was suddenly competition among the spendthrift princes to bring in the youngest and most modern Western girls they could find. They had to be beautiful, modern in outlook, but at the same time they must be both virgins and from bourgeois families. How costly this could prove was demonstrated by the example of the plump minister of education, Sheikh Jabir, whose eye lighted on the nineteen-year-old daughter of a north German hotelier. Jabir, sixty years old, had already been married thirty-one times, his last wife being a beautiful Lebanese. The German girl's mother insisted that before her daughter contemplated a union with the sheikh he must first rid himself of all his wives, and the ousting of the Lebanese alone cost him $3,000,000. The fair-haired Rhine-maiden eventually arrived in Kuwait to find wedding presents awaiting her of a stable of Arab horses (which she could not ride), twenty-three Cadillacs (which she could not drive), a load of precious jewellery (which she sent back to Germany to be reset), a rose marble palace on the Gulf (which she hated because it was too humid, and the Gulf smelled), and a husband whom she did not like. She was back in Germany within a year, divorced, but worth several million dollars more than when she had left.

The old ruler was indulgent about these peccadilloes. He had paid highly for a new young wife himself, a young girl he had seen in a swimming-pool during a sojourn in one of his palaces in the Lebanon. She came from Latakia, where girls were said to be wonderful with old men, and he was now past sixty. As for the reckless waste of money on palaces, on jerry-built schools and hospitals and roads: they could all be repaired, and they would all one day contribute to the image of Kuwait for which he was striving, rich, modern, prosperous. What matter if the image had cost more than it need have done? What matter if there had been corruption and waste? Had not thousands of Kuwaitis grown rich in the process? And what did money matter? Along the road into the desert the gas-flares lit the jabals like candlesticks on an altar, and out in the Gulf tankers queued up for Kuwait's inexhaustible supplies of oil. Why worry about expense, when there was so much oil to help pay for it?

'It is all part of the process of learning how to live with oil,' the ruler said.

But meantime, the citizens of Kuwait were beginning to grow accustomed to their heady wealth. Bedouin who had once been proud possessors of a brace of camels now carted their families around in brand-new American cars, and refused to send their sons to the ornate secondary and technical schools which the ruler had built. What would they learn there? To make furniture and mend cars and repair electrical circuits? But they were demeaning jobs, not respectable positions for the prosperous new Kuwaiti. The sons were sent off to learn more effete professions abroad. And into Kuwait to do the repair work and the dirty jobs came Arabs from other countries, hungry Iraqis and Yemenites and resentful Palestinians and Egyptians. They worked for the rich new welfare state but, except for their wages, drew no benefit from it.

They were the poor Arabs at the rich Arabs' table, and one day in 1956 they gave the profligate princes of Kuwait an ominous demonstration of just how bitter they felt about not being participants in the orgy.

18 · Blow-up

In July 1956, on the urging of John Foster Dulles, US secretary of state, the American and British governments withdrew their offers to finance the building of the new high dam on the Nile in Egypt. The high dam was the dream project of Egypt's new ruler, President Gamel Abdul Nasser, and this considerable slap in the face was meant to discipline the Egyptian and make him toe the line on certain onerous conditions which the Anglo-Americans were demanding in return for the loan. They misjudged their man. Nasser, riding securely on a surfing wave of Arab nationalism, replied by nationalizing the Anglo-French-owned Suez Canal. At the end of October, the British and French (having first made a secret agreement with the Israelis) moved an Anglo-French army into Egypt with the intention of taking the canal back by force.

Even in Victorian days, when Britain was the most powerful nation on earth, it would have been a foolhardy operation. In these twilight days of the British Empire, it was a suicidal move. The Americans, though sworn allies of the British, backed off at once and deprecated the use of force. Even the British ambassador in Cairo winced as he heard RAF planes overhead and saw British paratroopers coming down on Egyptian soil.

It would not be difficult to defeat the Egyptian forces, [he commented] but the difficulties would start after that. The Egyptians would organize guerilla warfare and it would be difficult for us to disengage against guerillas organized by Nasser or, if he had fallen, by his proclaimed successor. No government set up by the occupying forces would last. Only a government untainted by collaboration with the British could hold its position. If the British and French set up an international Canal management, they would find it difficult to withdraw their troops from the Canal and leave a foreign company, imposed by force, without their protection.[1]

From this unhappy, ill-thought-out, ill-prepared operation all manner of melancholy consequences were to flow in the months to come. British pretensions to world power would never be valid again. In the meantime, the repercussions from the political and military blunder at Suez were considerable all over the Middle East. But it was in Kuwait that they exploded most dangerously, and there it was not just against the British that they were aimed but against the Americans and the Kuwaiti ruling classes as well.

As has been suggested in the previous chapter, it could almost truly be said that every native-born citizen of Kuwait was a member of the Kuwaiti ruling classes. They were all privileged, even though some were more privileged than others. Soldiers in the small state defence forces were paid $150 a week, but were reluctant to volunteer because there was so much more money to be made with far less effort. Loans that would sound huge to British and American proletarians were freely available for every certified citizen with which to buy houses or businesses. A Kuwaiti had only to claim some obscure disease for one of his children and the whole family were flown to Europe, at government expense, to seek out expert medical treatment. But while ordinary (if ordinary is the word) Kuwaitis took it easy on government subsidies, while Kuwaiti businessmen garnered millions on government contracts, and while Kuwaiti princes poured out astronomical amounts in high living, the lesser Arabs who served at the rich Kuwaitis' tables looked on in envy and writhed in frustration over the inegalities of the situation. The Suez crisis and the bush-fire of anti-British anger which swept through the Middle East as a result gave them the opportunity for which they had been waiting to hit back at their effete and arrogant masters.

To begin with it seemed that the protest marches which began on the streets of Kuwait were the expected anti-British reactions for which the Kuwaitis had been prepared. The ruler and his advisers had anticipated that British business houses would be smashed up, the British flag burned, maybe even a few British subjects beaten-up and possibly killed. But that was not their affair. The British had asked for trouble, and now they would get it. So for the first forty-eight hours after the news of Suez began to come into Kuwait, and the fulminations of Cairo radio blared from the coffee-house loudspeakers, the local authorities shrugged their shoulders. The ruler, Sheikh Abdullah al Salim al Sabah, took the whole business with great calm.

Gawain Bell, the British political agent, went to tell him news of the British attack on Suez, and was filled with apprehension:

But I found the ruler perfectly composed. His strength is to know when to procrastinate—how to avoid becoming involved in the dissensions of Arab politics. His last words to me at this meeting were: 'My boy, go back to your files and don't worry!' He then sailed away twenty miles to Falaika Island and stayed there for three weeks, incommunicado with his own government or Her Majesty's Government![2]

By this time anti-British and pro-Nasser banners were flying all over Kuwait, and threatening crowds had gathered before the British Agency. Huge Kuwaiti cars drew up and whole families leaned out of the windows to watch the fun.

While they were doing so, one of Kuwait's notorious sand-storms came roaring in from the Arabian Desert and under its cover a number of saboteurs, in small groups, moved into Ahmadi, the oil town forty miles along the coast from Kuwait city and the headquarters of the Kuwait Oil Company. L. T. Jordan, the company's British general manager, was driving back through the storm from an inspection tour of the oil-gathering installations, when he saw a sudden flash of red through the brown murk ahead of him. Soon his car was skidding and sloshing through a morass of escaping oil, and black smoke from a blown-up pipeline had turned the sandstorm into a choking, impenetrable fog. Somehow, he managed to turn the car and make his way back to a sentry post, where he gave the alarm.

By that time nineteen other pipelines had gone up and were burning fiercely, their flames whipped higher by the gale-force wind. The saboteurs were even equipped for diving operations, and they slipped into the water alongside the submarine lines which carried oil to the tankers waiting out to sea. One of the lines blew up and began pouring thick crude into the Gulf. The order went out: 'Close down all wells. Seal off the tank-farm.' All over the desert, the wells stopped working and the fires by which the company flared off surplus gas guttered out. But the sabotaged lines burned on – and kept on burning for days to come.

This was not merely an anti-British demonstration now. It was a bid to shut down Kuwait's oil supply, and when news of what was happening in the desert reached Kuwait city, the authorities woke from their complacent apathy with a bound and proceeded to act with a speed and ruthlessness which made Anglo-American oilmen whistle with reluctant admiration.

The man in charge of Kuwait's internal security was a young relative of the ruler named Sheikh Abdullah Mubarrak. He had a pale, petulant face (he had inherited his complexion from a Circassian mother) which he adorned with twirly moustachios *à la Turque*, he gave sybaritic parties in one or other of the dozen palaces he owned, and he was said to have a solid gold field marshal's baton in his office safe. He ran his security forces as his personal bodyguard and paid them well over the wages granted to them by the state, and he forgave them anything if they kept law and order. Mubarrak now ordered them to the demonstrations still going on in front of the British political agency, and personally harangued the crowds, telling them they were victims of Communist agitators, and urging them to disperse. Those who did not go quickly enough he flailed with his stick and those whose appearance gave them away as Palestinians, Egyptians, Iraqis and Iranis, he ordered to be rounded up and taken away for questioning. Meantime, all Communist clubs and all cafes where foreign teachers and other workers gathered were being raided and hundreds of young men brought in.

They were beaten first and questioned afterwards. Mubarrak knew he did not have much time. Kuwait is a city which not only lives by oil. It runs on the derivatives of oil. Gas from the fields is piped into the city, and provides Kuwait's heat, light and power. If the oilwells were shut down for more than four days, the gas would be gone and life in Sheikh Abdullah's symbol of Arab progress would grind to a halt.* And police informers were coming in with reports that other bombs had been planted not only along the Kuwait Oil Company's pipe-

* As will be seen later in this narrative, all sorts of conditions (political and physical) can produce difficult moments for Kuwait energy supplies, which the local population is not likely to know much about, because no official announcements are made.

lines but also under those of the Neutral Zone operators, particularly along the lines leading from 'Gettysburg', J. Paul Getty's petroleum complex.

Thanks to Mubarrak's ruthlessness, the saboteurs talked. No fewer than one hundred time-bombs were found limpeted to the pipelines in the next twenty-four hours, and were defused by KOC explosive experts. One bomb went off at a wellhead in the desert south-west of Burgan, but was fortunately of such feeble quality that it failed to do more than superficial damage. But Kuwait had had a scare that it would not soon forget. As one of the sheikhs put it later:

'The saboteurs kept claiming that they were attacking British imperialism, but we were the sufferers – we Kuwaitis. It was our oil which was being sabotaged, our revenues which were being hit, our city which was threatened by a shut-down. There were a small number of Kuwaitis among the saboteurs, lazy, good-for-nothing dupes mostly. We sent them back to their fathers with the suggestion that a good beating might alter their foolish ideas, especially since their families would have been the first to suffer in the event of a complete sabotage of the oil-fields. But the ringleaders of the affair were more serious. Most of them were Palestinians and Egyptians, and this made it difficult. Both were martyrs of imperialism and the Zionist conspiracy, and therefore heroes even of those Kuwaitis whom they were trying to harm. We could not punish them. It would not have been popular. We sent them back to their jobs but henceforward we kept them under surveillance – ready for the time when we could more conveniently put them out of harm's way. With the rank and file of the militants, it was easier. They were nearly all Iraqis and Iranians. With them, we did not have to worry.'[3]

In mid-November 1956, as the Kuwait Oil Company began to limp back into normal production – though its output would be severely curtailed for many months to come – a purge of foreign labour began in the territory. Scores of Iraqis and Iranians were rounded up, bundled into trucks, and taken north to the Iraq frontier, where they were dumped into the desert to find their way to civilization as best they could. But with the more militant agitators, Mubarrak decided on even more brutal measures. The majority were severely beaten, but half a dozen of them were trussed and bastinadoed, and to rub home the lesson and *décourager les autres* the punishments were carried out in public, an action which infuriated the ruler when he heard about it at his retreat on Falaika Island. 'It will give Westerners a poor impression of our ideas of justice,' he said.

Despite his professed imperturbability, Sheikh Abdullah had been frightened by the outbreak of sabotage. The way in which the anti-British demonstrations had turned against the Kuwaitis themselves came as a surprise to him, but he was shrewd enough to realize their significance. The foreign workers were envious of the privileged members of the club, and as Kuwait prospered their envy would grow. He did not propose to appease them by transforming these poor foreigners into fellow members, with the same rights and privileges as the natives among whom they worked. That would be going too far.

But he did decide that Kuwait's 'honeymoon' was over, and it was time to call a halt to the reckless extravagances of his subordinate sheikhs. In the next few months, a regime of so-called 'austerity' was ushered into the territory. The ruler himself inaugurated it by cancelling the order for a new yacht and a private airplane.* Several new plans for sheikhly palaces were cancelled, and the most ornate of all, ordered by the ruler's son, was suddenly proclaimed a state guest-house. No Kuwaiti suffered from any shortage of cash in the days to come, but they became far more discreet in the ways they spent it.

And, to show that all Arabs were brothers, and that Kuwait was wholeheartedly on the side of the anti-imperialists, the ruler issued a proclamation expressing his detestation of the Anglo-French action at Suez, his hatred of Zionism, and his fervent admiration for President Nasser.

As one Palestinian put it: 'We were still travelling second class. But at least they were beginning to admit that we were travelling in the same boat. That was enough – for the time being.'

No euphemisms could conceal the fact that the Anglo-French operation against Egypt in 1956 was an abject disaster, morally, militarily, politically. It ruined the career of the British prime minister, Anthony Eden, who had ordered the invasion. It con-

* He 'borrowed' one instead from the newly-formed Kuwait Airways, which in turn went to the Kuwait Oil Company for the money with which to buy a replacement.

vinced the world – and the Middle East in particular – that the sap was running out of the British imperial oak, and those who had sheltered in its shade began to look for better sources of protection. Others began to carve their names on its trunk.

The repercussions all over the Middle East were immediate, and so far as the oil business was concerned it was inevitable that any reprisals against British interests were bound to involve US companies. With the exception of Saudi Arabia, where the concessions were wholly American-owned, every major British operation had US participation to a greater or lesser degree. Saudi Arabia continued business as usual with Aramco while at the same time breaking off diplomatic relations with Britain, but elsewhere the Arabs hit back at the Suez invaders in the only way they knew how: by blowing up pipelines in Iraq and Syria, by going on strike in the oilfields, and by blocking the exports of oil to Europe.[4] American oil exports from the Middle East were held up equally with those of the British, and much of the contumely spilt over on to their shoulders too.

No less than in the Arab countries, anti-British feeling was strong in Iran, where the memories of past intervention by British troops were still bitterly remembered. The resentment was directed in this case against the new oil consortium, in which, of course, the Americans now split with the British the major proportion of the shares. The consortium had just begun to assume its share of the world oil markets, and the oilfields were busy and thriving, even though Egypt's blockage of the Suez Canal after the British invasion forced the combine, for a time, to redirect its shipments.

But the consortium was not popular. From the shah downwards, most Iranians considered that the US negotiators had driven far too hard a bargain, and that only Iran's desperate need to get the oil flowing again had forced them to accept the terms. Now resentment welled up and mingled with anger over Britain's high-handed actions in Egypt; and had not the ink only just dried on the consortium agreement, there might well have been a campaign to get it amended. But it was too soon. Changes would have to come later. In the meantime, the Iranians set about making sure that they would never have to accept such onerous conditions again for the exploitation of their own oil. In the summer of 1957, the Majlis in Teheran passed into law the Petroleum Act. It was the innocuous name for revolutionary new legislation – and it brought Enrico Mattei of Italy back on to the international oil scene.

Enrico Mattei had been busy in the two years since the big Anglo-American oil combines had snubbed and humiliated him and, in his view, cheated him out of a share of the Iranian consortium. Most of his time had been devoted to devising ways and means of solving Italy's chronic power problem – its almost total reliance for its petroleum supplies on the international oil companies. As president of Ente Nazionale Idrocarburi (ENI), the Italian national oil company, he was more than ever eager and determined to find oil supplies independently of the cartel, and if he could do so in such circumstances that he made life difficult for the Anglo-Americans, so much the better. Mattei had been an opponent of Benito Mussolini but he shared the erstwhile dictator's flair for dramatics, and the sense of insult he felt at the way in which he had been snubbed by the consortium was genuine. He was a proud, even arrogant, man and he would never forgive the Anglo-Americans; and any arrangement which he could make which would do them harm would be all the more rewarding, so far as he was concerned.

The post-Suez climate of downright anti-British and subdued anti-American feeling throughout the Middle East was just what he needed for his purposes. It produced a revulsion of feeling against the big oil combines, a conviction that the governments and people of the area were being cheated by the foreign exploiters in their midst, and a widespread determination, as Walter Lippman wrote about that time, 'that concession agreements for oil production and exploration in new areas must be based on different principles from those established by the big international companies.'[5]

During the Suez crisis and in the uneasy months which followed, Enrico Mattei spent a great deal of his time both in Cairo and in Teheran. In the Iranian capital he had several talks with the shah, and with the petroleum minister, Dr Amini, and he was present in the Majlis when the Petroleum Act was passed into law. As a weapon for hitting the oil combines where it hurt most it could hardly have been better devised, for it introduced something against which the Anglo-American companies had been fighting for most of their lives: a scheme for a joint enterprise in which the oil company and the oil producing country participated on equal terms.

This was very different from the so-called fifty-fifty arrangement which Aramco had worked out with the government of Saudi Arabia, an arrangement which they often loftily referred to as a 'partnership'. As the oil historian and consultant, George W. Stocking, has remarked:

> The 50-50 profit-sharing agreement . . . is about as much a true partnership as is the relationship between the United States government and the thousands of corporations that it subjects to corporate income taxes.[6]

Aramco paid the Saudi government fifty per cent of its earnings in income taxes, but ran its company as a separate entity on behalf of its four chief US holding companies. The Saudi government had one director (and later two) on the Aramco board, but without voting rights or direct influence on policy. That was the way it was and that was the way the Anglo-Americans wanted to keep it.

On the other hand, the proposals which Mattei outlined to the Iranians, and were subsequently inculcated into the Petroleum Act of 1957, envisioned a joint venture in which the oil-producing country was involved from the beginning, although at no financial risk. The first of these agreements was signed between the Italian subsidiary, Azienda Generale Italiana Petroli (AGIP), and the government-owned National Iranian Oil Company (NIOC) almost immediately after the passing of the Petroleum Act into law, and the agreement was approved by the Majlis on 24 August 1957. It was described by the Iranians as the first result of 'the most progressive legislation in any Middle Eastern country', and by an American oilman as 'a kick in our corporate guts'. AGIP and NIOC formed a company between them, SIRIP,* which would have exclusive exploration rights over 5,600 square kilometers of territory north of the Persian Gulf, projecting out along the continental shelf; 11,300 square kilometers on the eastern slopes of the Zagros mountains; and 6,000 square kilometers along the Persian side of the Gulf of Oman.

The Anglo-Americans whistled with astonishment when they read of the obligations which Mattei had accepted. The

* Iranian Italian Oil Company.

chairman of the board of the new company would be Iranian
and the directors would be Iranian and Italian, in equal
numbers. All the labour and as many of the technical jobs as
could be filled by qualified personnel would be assigned to
Iranians, and a training school for others started at once. The
Italians agreed to spend $22,000,000 looking for oil, $6,000,000
within the first four years and $16,000,000 within the next
eight years. The moment one commercial well had been
discovered, the Italians were pledged to go on looking for new
ones until the whole $22,000,000 was expended. Only when
wells were ready to go into production would the Iranians
come in to share the costs. Thereafter, both the Italians and the
Iranians would take 50 per cent each of the oil, and would share
the profits between them. But since SIRIP would be paying a
50 per cent tax on their share, and the Iranian company NIOC
would be getting 50 per cent of the remaining profits, the
Iranian Government would in fact be making 75 per cent on
the deal.

When the oil industry spokesmen had recovered their
breath, they let out a whoop of astonishment and derision.
Only a man desperate for oil at any price, and motivated by a
vendetta against the oil combines, could possibly have accepted
such onerous terms, they implied. The more extreme among
them sharply suggested that Mattei's only aim was to sabotage
the holdings of the big companies in the Middle East. To that
he replied:

It should not be necessary for me to affirm my country's right
to insure its own independent oil supplies. The supremacy of the
international oil cartel is not taboo and Italy does not intend to
respect it. Besides, today the cartel is everywhere being broken
down. Through the State organization of which I have the honour
to be president, Italy can now enter the field itself to obtain its
oil supplies . . . I have no doubt that Italy's agreement with Iran is
not only in our interest but in the general interest also.[7]

Nonsense, replied the big companies. On such terms, no one
would be willing to come in and make a deal with a producing
country. The risks were too great and the price was too high.

But three months later, the Iranian government announced
that further area of the country was being offered as a concession under the new Petroleum Act, on a joint-venture basis.
No member of the Anglo-American consortium bothered to

apply, and their spokesmen sceptically suggested that no one else would, either. But he was wrong. Fifty-seven companies from nine different countries applied for questionnaires – which they had to fill in if they were to be considered – and Iranian experts who subsequently monitored their answers deemed eleven companies 100 per cent competent, financially and otherwise, to embark on exploration under the terms of the act, and ten others between 10 and 75 per cent competent. All the applicants were eager enough to have paid $2,700 each for the documentation.

One was chosen from among them. To the dismay of the consortium members, this time the winner was not a maverick of the Mattei type but a big, respectable American oil company, Standard Oil of Indiana, bidding in the name of a subsidiary, Pan American International.

A week or two later, an oil company vice-president on his way home on leave, stayed overnight in a Rome hotel and saw Enrico Mattei walking through the main lobby.

'We aren't buried yet,' he said, sourly, 'but there goes the man who has driven the first nail in our coffin.'

The Middle East oil business would never be the same again. Someone later estimated that snubbing Mattei had cost the Anglo-American oil cartel a billion dollars in revenue. But it had done more than that, as the companies would very soon discover.

19 · Enter the Japanese

Enrico Mattei had really started something. All over the Middle East the sheikhs were beginning to repeat the words 'joint ventures' as if they were holy writ, and the major oil companies realized, with sinking feelings, that the interloper from Italy had changed everything. If they had been taken aback by the liberal terms he had conceded to the Iranian government, they were alarmed and dismayed by what followed. For the conditions Pan American International accepted for its joint venture with the Iranians were even more expensive. The company made an immediate payment of $25,000,000 just to get the concession, whereas Mattei had paid nothing. It agreed to spend $82,000,000 on exploring and developing the concession, compared with Mattei's agreed outlay of $22,000,000. Otherwise the conditions were similar.

One thing was certain. The days of the old type of concession-agreement were over. True, most of the concessions held by the Anglo-American major companies had several decades to run,* but for how long would the Arab and Iranian governments, operating the new 'joint ventures' so much more profitably with their foreign partners, continue to accept the old-style monopolist companies operating beside them? There were difficult times ahead, and the name of Enrico Mattei was not one to be mentioned in the board-rooms of the major oil companies in London, the Hague and New York.

In Saudi Arabia there was one man who had followed the Italo-Iranian negotiations with fascinated interest, and he saw in the adoption by his country of Mattei's new-style partnerships an ideal weapon for chipping away the foundations of

---

* The earliest to mature would be the consortium in Iran, which was due to expire in 1980. But some – Kuwait, for instance – extended beyond the end of the century.

the all-powerful American monopoly, Aramco. His name was Abdullah Tariki, and he had no love for the United States, Aramco, or any other big capitalist organization.

In 1957 Tariki was thirty-two years old. A son of a government official, he had been sent to Cairo university at the end of World War II, and had there taken a degree in geology, at the same time making friendships and mopping up ideological ideas among the young anti-British, anti-monarchy Egyptians who would, a few years later, be the most fervent supporters of Gamel Abdul Nasser. From Cairo he had gone on to the university of Texas, where, in 1947, he had taken a master's degree in petroleum geology, and afterwards spent a year on a training course with the Texas Company (one of the chief shareholders of Aramco). There, although he will not talk about it, something seems to have happened to him that soured him against Americans. Offered a job with Aramco when he returned to Saudi Arabia, he turned it down. Instead he enrolled as a minor official in the Bureau of Mines, and found himself acting as a liaison officer with Aramco at their headquarters in Dhahran.

There too he appears to have suffered some sort of traumatic experience which left him with something more than an antipathy towards the American company. In its treatment of its own Saudi technicians and employees, Aramco's American direction is impeccable. The company's policy is to help them in every way possible, in educating themselves and their sons, in loaning them money for houses, in training them for promotion; and once they begin to climb the technical ladder, they are accepted and befriended. But with minor Saudi officials employed by the government, Aramco officials could be both curt and rough. The courtesies were reserved for the sheikhs and the ministers, and less time was taken with small fry. It may have been a verbal tussle with some exuberant Texan, or, as some say, because one Aramco manager once kept Tariki standing for two hours in the hot sun while he kept cool in an air-conditioned car, but whatever it was, Aramco was not Tariki's favourite company. By 1957, he had become director general of petroleum and mineral resources to the Saudi government, but the rise in his station did not mellow him. Among his close associates he talked bitterly of 'American economic imperialism' and argued persuasively – he was a brilliant speaker – for the need to introduce new elements into the exploitation of Saudi oil in order to cut down Aramco's monopoly.

There was one promising area of Saudi Arabia which was well outside the vast concessions which Aramco controlled, and that was on the continental shelf offshore from the Saudi-Kuwait Neutral Zone.* Some of the major companies and a large number of independents had already made tentative bids. Among them was a company called the Japanese Petroleum Trading Company, composed of a number of diverse banking, electronic, insurance and manufacturing groups (Mitsui and Mitsubishi among them). Its chairman, Mr Taro Yamashita, had already had talks with a number of ministers and officials in Jiddah, and to Abdullah Tariki he was more than welcome. He was afterwards to say: 'Yamashita was our Mattei.' Like Mattei, Yamashita represented a nation without its own oil resources, up till now dependent on supplies from the major Western oil combines. Like Mattei, he would be in direct competition with the Anglo-American cartel. Best of all, from Tariki's point of view, he was not an American.

From several preliminary conversations with Mr Yamashita and his advisers, Tariki worked out a draft agreement which he was convinced would give Saudi Arabia its first 'joint venture' and would include all the advantages which Iran had secured, and more. But he anticipated trouble when he went to Riyadh to present his plan to King Saud and to his younger brother, Crown Prince Faisal, who was also prime minister. He knew how eager were the other applicants for possession of the new concession, and such was the climate of corruption at the Saudi court that he feared that the Americans would already have bought their way in, and, what is more, done so on much less rigid terms than he had secured from the Japanese.

To his surprise, neither the king nor the prime minister raised any objections to the agreement, and made no demur at the entry of the Japanese into the Middle East oil business. There was, of course, no reason at all why they should have opposed the agreement, and every reason why Tariki should have been congratulated for having driven such a hard bargain with the newcomers. But their sheer lack of comment or

* Where Jean Paul Getty and Aminoil between them had developed highly profitable production.

criticism puzzled him, for he knew the ways of the royal household only too well.

When he returned to Jiddah and told Mr Yamashita that all was well, the Japanese indicated that he had known all along that there would be no royal objections. From his briefcase he extracted a document and passed it across to Tariki, who read it in growing stupefaction.

It was an agreement, signed the previous June, which stated in part:

Whereas the Japan Petroleum Trading Company Ltd desires to obtain a concession from the Saudi Arabian Government that gives her the right to search and prospect oil in the Saudi Arabian territory. Mr Taro Yamashita, president of the Japan Petroleum Trading Company Ltd and His Excellency Kamal Adham have agreed the following:

1 That Mr Hassan Khalifa agrees to cooperate with Mr Taro Yamashita to promote the exploitation plan under the supervision of His Excellency Kamal Adham.

2 That the contract regarding the said concession is to be concluded between Mr Taro Yamashita, as president of the Japan Petroleum Trading Company Ltd and the Saudi Arabian Government . . .

3 That in case the above mentioned concession is granted by the Saudi Arabian Government . . . Mr Taro Yamashita . . . undertakes to pay His Excellency Kamal Adham two per cent (2%) yearly of the share of the total net profit accruing from the said concession . . .

4 That Mr Taro Yamashita . . . will pay Two Hundred and Fifty Thousand U.S. dollars (U.S. $250,000) to His Excellency Kamal Adham within one month after the conclusion of said contract . . . and Seven Hundred and Fifty Thousand U.S. dollars (U.S. $750,000) upon sufficient amount of petroleum be [sic] found as to permit sound business enterprise. . .[1]

His Excellency Kamal Adham was Crown Prince Faisal's brother-in-law.

Still rocking on his heels from the impact of the document, Tariki asked the Japanese why, with such royal allies in his pocket, he had accepted such onerous terms for the concession. Mr Yamashita smiled. One of these days, he said, the big American companies and their old-style concession agreements would be in trouble, and he did not wish the Japanese to be associated with them when that happened. 'Joint ventures' were the way things would henceforth be arranged, and Japan would be content with them.

'It has also to be remembered,' he said, 'that we have only 50 per cent of the concession so far. We now have to obtain the other 50 per cent from the Kuwait government. And there we do not have a good friend like His Excellency Kamal Adham. At least we do not have one yet,' he added. 'It is as well to demonstrate in advance to the Kuwaitis that we are prepared to go very far in order to secure a complete agreement.'

In May 1958, the Japanese Petroleum Trading Company Ltd successfully outbid all other contenders – and the competition from the US and other countries was fierce – for the Kuwait 50 per cent rights of the offshore Neutral Zone concession, giving it 100 per cent of control of all rights to drill in the area.

The terms agreed to by the Japanese were in both [Saudi and Kuwait] cases unprecedentedly rigorous [writes Stephen Longrigg]. In the concession given by Kuwait the full concession period, including exploration, was 44½ years only, and in the Saudi concession two years more : and in both the right of renewal was only partially assured, and not at all if a rival Kuwaiti or Saudi candidate were in the field. In each agreement, a heavy surface rental was payable from the date of signature, and was to increase with the passing years if oil was found.[2]

In the agreement Tariki had drawn up, the Saudi Arabian government received 56 per cent of any profits made (the Kuwaitis got 57 per cent). These profits were to be calculated not just on sales of oil from the field, but on any stage in the sales pattern and in any country. Moreover, the contract stipulated that a) no oil could be sold to 'enemies of the Arabs'; b) the Saudi government would have a substantial shareholding in the Japanese company and two directors on the board, who would supervise operations and policy; and c) in the event of difficulties, the Japanese would agree not to appeal to their own government for protection.

Even compared with the Mattei agreement, they were stiff terms, and once more the old-style companies prophesied that the price was too high and the Japanese would run out of money before they found any oil. But, 'undeterred by these provisions, and driven forward by their country's known and serious need for oil',[3] the Japanese went to work almost at once. Under the name of the Arabian Oil Company they assembled a mixed Saudi-Kuwaiti labour force and a cadre of Japanese and other foreign petroleum experts at a base on the shores of the Neutral Zone, and brought out a mobile drilling platform from Texas. Disaster struck on the afternoon of 3 April 1958. The mobile platform, moored eight miles from the shore, had been drilling in 100 feet of water. At 1,500 feet below the sea bed the drill tore its way into a large pocket of natural gas, which bubbled to the surface in a sudden eruption and ignited on contact with engines aboard the platform. The subsequent fire raged for eleven days before it could be smothered out by experts flown in from Texas. In the meantime, the drilling platform had been beached, badly damaged.

It was an expensive start to the new venture, and the big companies, watching from the sidelines, predicted that even the Japanese would now see that the terms were 'unworkably onerous' and would get out before their reserves were burned up along with their drilling platform. But the Japanese were not giving up that easily. They sent for new equipment, and began again, a few miles inshore from the gas pocket.

On 1 January 1960, the Arabian Oil Company brought in oil with its first well – and it produced 6,000 barrels a day at 5,000 feet. Two years later, the company had 34 wells operating and by 1964 they were producing 240,000 barrels a day. They were in business – the oil business. And Mr Taro Yamashita was proudly reporting to his shareholders that Japanese oil ventures in the Middle East had 'a high future full of big dreams and hopes'.[4]

Nobody was happier about it than Abdullah Tariki. Not that he was satisfied. His campaign against the big oil combines had only just begun.

'The trouble with a lot of guys in the big companies,' said Howard W. Page, a vice-president of Standard of New Jersey, 'was that they were always talking about the sanctity of contracts. What they couldn't see was that in the light of what was happening in the Middle East, a lot of these agreements no longer made sense. In other words, if we had started out afresh and worked out new contracts we would never have insisted on many of the things that were in the agreements. But there were – and still are – a lot of people in our industry who insist that a contract is a contract, and that the signatories have got to live up to it. Actually, in contracts between companies, when things get out of whack, we usually agree to adjustments in the interests of good business. But they wouldn't think of it as far as contracts with countries were concerned. So when we got into negotiations with Iraq, some companies didn't agree to anything – all on the basis that contracts were unbreakable.'[5]

The Iraq negotiations to which Page was referring were those of 1951, when, thanks to his patient and skilful advice, a fifty-fifty agreement of the type which Aramco had signed with Saudi Arabia was accepted by both parties. But by 1958, this type of artificial 'partnership' had soured on the Iraqis, and they were demanding a new deal. Thanks to the closing of the pipeline connecting the Iraq oilfields with the Mediterranean – it had been blown up by the Syrians during the Suez crisis – shipments of petroleum had gone down drastically and Iraq's revenues with them. The country was badly in need of money. Since the Iraq Petroleum Company (IPC) was mainly British-owned and British-staffed, it had become the target of much of the anti-British feeling which swept the country after the Anglo-French attack on Egypt. To more militant Iraqis not even 'joint venture' agreements now had any appeal. They were shouting for outright nationalization.

It was in this climate of seething, emotional discontent that IPC's managing director, G. H. Herridge, arrived in Baghdad on 4 July 1958, to begin talks about the concession agreements. The smell of revolution was in the air, but the IPC delegates appear to have mistaken it for the reek of petroleum and blithely negotiated as if Suez had never happened. Eight days after their arrival, the British negotiators left again for London to consult head office, confident smiles on their faces, breathing optimistic words about 'friendly atmosphere' and 'mutual goodwill'. They indicated that they had agreed to surrender certain parts of their concession and to double their production of oil within the next two years. These were gestures hardly likely to appease even the pro-British clique which ruled Iraq at the time, for Nuri Said, the Iraqi strong man, needed much more liberal gestures to quell the rising tide of discontent. Forty-eight hours after the British left Baghdad, the pro-British clique no longer existed.

On 14 July 1958, Iraqi troops burst into the grounds of the royal palace in Baghdad and demanded the surrender of the young King Faisal and his uncle, Prince Abdulillah, who was acting as regent. When the two men, accompanied by Abdulillah's mother and the king's two sisters advanced towards the troops across the palace lawn, they were shot. Abdulillah, a petulant, perverse and arrogant character, was cordially hated throughout Iraq, and his body was thrown to the crowd outside, where it was joyfully seized for mutilation and 'dragging'. Iraqi mobs have two favourite spectator-sports during times of crisis and revolutions: one is public hangings in the main square of the capital, the other tying the body of a designated public enemy to the back of a truck and dragging it through the streets. Abdulillah was not the only one to be 'dragged' in the next few hours.

Nuri Said, Iraq's prime minister, friend of Britain, inflexible diehard, was the main target of the army revolutionaries, but he managed to escape across the River Tigris, where he hid in the house of a friend. A servant, thought to be faithful and devoted, was paid a large sum of money to reveal his whereabouts, and he was trapped while trying to slip away from the friend's house disguised as a woman. He too was shot and handed over to the mob for the grisly ceremonials.

Meanwhile, the crowd was going on the rampage all over the city. A number of foreigners, including three Americans, were dragged from their hotel and taken around the capital in trucks until the impatient mob swarmed upon them, beat them to death, and 'dragged' them too. One of the murdered Americans was named Eugene Burns, of Sausalito, California, and the rebels afterwards claimed that he was a member of the CIA, masquerading as a fund-dispenser for a children's charitable organization. If so, he must have been singularly inefficient, for it turned out later that both the CIA and the British secret service were taken by surprise when the rebellion broke out.

The mob next turned its attention to the British embassy, where it shot one of the staff, robbed the rest of their rings and watches, and burned the building to the ground. By nightfall the country was in the hands of the rebels, a group headed by an army officer, Abdul Karim Kassem, who had always been accepted by Abdulillah and Nuri Said as one of their closest friends. Wrote Sir Humphrey Trevelyan, who took over as

British ambassador when the smoke of burning buildings and the stench of dead bodies had been cleared away:

The Hashemite monarchy, established by the British nearly forty years before on the ruins of the Turkish Empire and bound by alliance to the British, was overthrown. The romance of early Anglo-Arabism, the special relationship born in archaeology and the coincidence of wartime interests, pictured in Allenby's capture of Jerusalem and the contrived entry of Faisal I into Damascus, incarnate in those more than life-size figures of the romantic period, Lawrence, Gertrude Bell, Percy Cox, was finally shattered in the ruins of the British Embassy in Baghdad. The Hashemite-British alliance had outlived its time, while Nuri Said, the prime minister who had seen it all from the beginning, grown old, rigid and careless of security, had failed to adapt himself to the political needs of the time. The crash came, and in one day the old links with Iraq were severed, never to be restored.[6]

It might have been thought that conditions were now too fraught and unfriendly to warrant any resumption of negotiations between the Iraqis and IPC over the oil concessions. A victorious rebel leader, eager to conserve the enthusiasm of the mob, was hardly likely to be in the mood for reasonable negotiations. But suddenly the US government came to the aid of the British and proceeded to do in the Lebanon what it had so strongly deprecated when the British had done it in Egypt two years earlier: it landed American troops. True, President Eisenhower announced that the marines had landed 'at the request' of the Lebanese government, disturbed at the troubled revolutionary situation in the Middle East, but there were not many Arabs who accepted that explanation. One of the most surprised of them, indeed, was the Lebanese prime minister when he heard the president's broadcast announcing that he had asked for help. But if the Arab world was mortified by the American action, it was in no position to do anything about it – at the moment, anyway. Once more they would remember the old Arab saying: 'The hand you cannot bite, kiss it.' Kassem kissed it, reluctantly. The State Department sent under-secretary William Rountree to see him in Baghdad, and though the American was hooted on the streets by demonstrators, the Iraqi leader received him cordially enough. On 18 July he announced in a radio message:

In view of the importance of oil for the wealth and the world economy, the Iraqi Government announces its wish for continued production and export of oil to world markets. It also upholds its obligations to all parties concerned. The Government has taken the necessary measures to preserve the oil fields and oil installations. It hopes the parties concerned will respond to the attitude taken by it toward the development of this vital source of wealth.[7]

Spokesmen for the new regime were soon assuring everyone that there was no intention of nationalizing the IPC concessions. Said Colonel Abdul Faik at the Iraq embassy in London:

It is not the intention of the Republic to think about this, because we believe that if the oil continues to flow to the usual markets, it will be for the benefit of both parties—you get your oil and we get our pounds.[8]

In January 1959, a new delegation arrived in Baghdad from IPC to start talks with the Iraqi government. This time it was led by IPC's new chairman, Lord Monckton, no oil man but an amiable lawyer famed for his skill as a negotiator.* But more amiability than skill emerged from the meetings. Lord Monckton declared, upon his return to London: 'I do not remember any discussions conducted in a more amicable spirit, and they were continually saying that they wanted to work with us to our mutual advantage.'[9]

But work with them on what, and in what circumstances? The Iraqis announced that IPC would immediately get its expansion programme under way, and would double the export of crude oil from Iraq by 1962. The company would also begin a construction plan for the erection of oil refineries inside the country. The following day, IPC called in the press and read out to them a prepared statement saying that the expansion programme would be subject to 'world market conditions', which, the spokesman reminded his listeners, were not very promising at the moment. As to oil refineries, IPC had no intention of building one in Iraq.[10]

It was a moment when IPC would have done well to make a gesture, and a generous one, to meet the political realities in

---

* He acted as go-between during the Abdication Crisis in the discussions between the then King Edward VIII (later Duke of Windsor) and Prime Minister Stanley Baldwin. He was minister delegate to the Middle East during World War II.

the new Iraq. Company spokesmen and industry-oriented historians have since maintained that it would not have done any good, that Kassem was to prove a vacillating trickster whose word could not be trusted, and who ended up with an assassin's bullet in him, anyway. On the other hand, some of the sager petroleum experts dispute this.

For quite a number of years, [says Howard Page] I was looked on as a dangerous liberal for agreeing to study situations, particularly where you could analyse them and say, 'Well, if you were starting all over again, this is what you would have to give.' So long as these guys (the producing countries) didn't have control of us, we weren't prepared to give up anything worth a damn. I was looked on as a dangerous character for suggesting that we should.[11]

And that was the trouble. If Kassem was to prove a trickster it is possible that IPC helped to make him so. There was a world surplus of petroleum and too much fuel was competing for too few markets. Oil producers in the United States had beefed up their output during the Suez crisis and were happily selling their petroleum unworried by foreign competition, because Congress had put a quota on imports of foreign oil. Russia had begun flooding the markets in Europe and Asia with Soviet oil at cut prices. The new independents like J. Paul Getty and Continental were beginning to market their products, and were, so far, under no control by the big combines. IPC had plenty of other places in the world where it could buy oil for its markets, and so did the companies who made up its membership. There was all the fuel they wanted in Saudi Arabia, Qatar, Abu Dhabi, Kuwait, and South America.

So each demand that Kassem made of IPC – and in the light of present-day developments, they were not all that formidable – was met not by concessions but with a deliberate, if unpublicized, tightening of the screw. And if Howard Page was against this type of pressure, he was one of the few. For there was no doubt now that IPC's confrontation with the Kassem government was being watched by all the major Anglo-American companies, and a hard line supported by all of them.

They realized that Iraq could scarcely have been unmindful of Iran's unfortunate experience with nationalization, which had shut it off from the world's market and eventually triggered the coup d'état which overthrew the Mossadeq regime [writes George W.

Stocking]. The Iran crisis occurred when surplus capacity was far less and the oil industry had readily met it. . . Perhaps more important to the companies was the realization that any major concession they made to Iraq would be promptly demanded by other Middle East governments, constantly guided in their relations with their concessionaires by the 'most favoured nation' principle. The companies were not ready apparently to upset entirely the pattern of Middle East concessions, already greatly weakened by the advent of the newcomers.[12]

So pressure was the order of the day for the IPC negotiators, with combined industry approval, British and American. The most obvious way to make its strength felt was for IPC to deprive the Iraqis of the increase in revenue upon which Kassem had been banking after the British company had announced its expansion plans. Far from doubling its output, the company quietly ordered its wellhead managers to slow down the flow. In the next three years, while negotiations were proceeding, IPC's production of Iraq oil increased by between half and one per cent. In the same period Iran's production went up by 12 per cent, Saudi Arabia by 9 per cent and Kuwait by 11.5 per cent. In one particularly critical month of IPC-Iraqi confrontation, the company's production dropped by 30 per cent.

To stress the international character and solidarity of the negotiators, IPC's managing director, Herridge, was joined in Baghdad by F. J. Stephens of Shell and H. W. Fisher, a vice-president of Standard of New Jersey. But they had been instructed to keep the existing concession contracts in mind as they bargained, and anything beyond the agreements they had made in 1951 – with minor adjustments – were anathema to them. From Fisher's point of view, to echo Howard Page, a contract was a contract. When Iraqi leaders asked them for a gesture that would give the Iraqi people a say in the exploitation of their rights, Fisher replied: 'This is a commercial arrangement between the two parties. Much as we feel for the hardships of the Iraqi country and people, we should . . . adhere to the commercial basis. . . .'[13]

It was this attitude of boneheaded commercial inflexibility which gradually drove the Iraqi leader into the extreme positions which eventually led to a breakdown. As one of his ministers put it:

The other day, and for the umpteenth time, I was trying to explain to the company representatives just how negative and absurd their attitude was. You are making strenuous efforts, I told them, to help the underdeveloped countries by granting them long-term loans and technical aid. You are doing this, it seems, to raise their living standards and to shelter them from subversive doctrines. We don't ask this much from you for Iraq. . . . Just give us our due. Try at least to behave in good faith. You take a malicious pleasure in misleading us, and in depriving us of our most legitimate rights. In the minds of the people, all this may finally rebound against you and against the principles of what is still called 'The Free World.'[14]

What were the demands which Kassem was asking of IPC? They would certainly have necessitated a complete reconstruction of the concession agreement, but seen from the viewpoint of the 1970s – when most of the demands have been conceded or are being discussed – they do not strike terror in the heart even of the most diehard oilman. They can be summed up as follows:

1 The government pointed out that although it had the right to appoint two directors to the board of IPC and its subsidiary companies operating in Iraq, in practice the board had only limited functions and policy was controlled by executive directors, operating in London. The government demanded that one of the executive directors should henceforth be an Iraqi appointee.

2 Desiring to enjoy the profits of the integrated operation of IPC oil business, the government demanded that the company give priority to the shipment of crude oil in an Iraqi tanker fleet that the government was planning.

3 All over the oilfields, IPC was flaring away surplus gas. The government demanded that IPC either pay the government for wasting it in this way, or arrange to have it piped to government depots.

4 The government pointed out that under the terms of the concession contract, whenever an issue of IPC shares was offered to the general public, Iraqis in Iraq were to be given preference in being allowed to buy at least 20 per cent of the issue. But it had since been discovered that this privilege was illusory, because IPC was a private and not a public company, and therefore shares were never put up for sale and neither Iraqi individuals nor the Iraq government were able to purchase stock. The government now demanded the right to purchase a 20 per cent participation in IPC.

Clauses 2 and 3 of the Iraqi demands were not exactly vital to IPC's monopoly, and the company were prepared to make mollifying gestures in regard to them. Yes, they would advise the Iraqis about how to establish a tanker fleet, and would use their ships for transporting Iraqi oil, once the fleet was established (which, they suspected, would not be for a long, long time). And yes, they would make arrangements for the bulk of the gas now being flared to be used by the government for fuelling and powering its cities.

But clauses 1 and 4 were much more fundamental. To have given way on either would have been to admit Iraqis into the board-room of IPC, with a say in policy, prices and output. That they would never accept.

Soon both sides were accusing each other of trickery or intransigence. The more IPC procrastinated, and cut down its production, the more the emotional Iraqi leader was driven into an extreme position. The less IPC was willing to give, the more he had to ask – to show his people that he could not be crushed and humiliated by these foreigners. When the British company announced that it was, for the time being, closing down operations in the potentially rich oilfield of Rumeila, in southern Iraq – part of the vast concession which IPC had held since 1925 – Kassem's response was swift.

On 11 December 1961, he announced that – under a new act called Law 80 – he was withdrawing all but 1,937 square kilometers of IPC's concession. This was 0.5 per cent of the total area, and represented the territory in which most of the company's wells were in active operation. The remaining 385,700 square kilometers would revert to the government and would be offered to new concessionaires.

IPC promptly referred the confiscation to international arbitration, and both the British and US governments sent notes to Kassem urging him to arbitrate. He brushed them aside.

'Why should IPC continue to hold such vast areas of our country?' he asked. 'The company is only utilizing three per cent of it, and keeping the other companies from exploiting the rest to the benefit of our people.'

The confiscated areas were offered to world oil companies for immediate exploitation. But IPC was not worried. It was calmly confident that the Anglo-American cartel would hold firm, and no one would bid. The company was quite right. The international oil-front held, and the confiscated areas lay fallow for the next seven years. No one would touch them.

But IPC would live to regret its intransigence. In 1963 Kassem was shot down by assassins, and his severed head displayed on Iraqi TV screens, to cries of: 'The pro-British puppet is dead! Long live the revolution!' The men who succeeded him were even more fanatic, even more demanding, and by this time the world oil situation had drastically changed.

Outside Iraq, another kind of revolution had taken place – and so far as the big oil companies were concerned, it was of a much more threatening nature than the exigencies of General Kassem.

20 · Over a Barrel

In the summer of 1960, Standard Oil of New Jersey did something which united the oil countries of the Middle East as nothing had ever done before, not even their mutual hostility towards the State of Israel. Jersey cut the posted price of its Middle East oil by fourteen cents a barrel, and news of the company's unilateral action reverberated through the Arab world like the shots at Sarajevo. Soon the other major companies began cutting their prices, too. From Baghdad to Riyadh, and from Teheran to Kuwait, the sheikhs and the shah were hit in that most vulnerable and painful part of them, their pockets, and they were loud in their cries of distress and anger. More important than that, they decided to take joint action against the cuts, and thus save the posted price system upon which Middle East oil production was based.

This is probably the place to explain what the oil industry means when it talks about posted prices. It will have become apparent in the course of this narrative that – except for the new independents – all the oil companies operating in the Arab countries and Iran were (and still are) subsidiaries of the Anglo-American majors. The Iran consortium is owned by BP, a quintet of American oil giants,* Compagnie Française des Petroles, and Shell. IPC of Iraq belongs to BP, Shell, the French, Standard of New Jersey and Mobil. Aramco of Saudi Arabia is divided between Standard of California, Standard of New Jersey, Texaco and Mobil. Kuwait Oil Company is a condominium of BP and Gulf. And so on. These subsidiaries are producers of oil and have no hand in marketing it outside the producing countries. They sell their oil to their parent companies, whose responsibility it then is to sell it to the world. In the old days, the subsidiaries would sell crude oil to

* Standard of New Jersey, Mobil, Standard of California, Gulf and Texaco.

231

their parent companies on low-price contracts, and pay a royalty based on that low price to the governments of the countries in which they operated. Then their parent companies went on to sell this bargain-rate crude at the prevailing world prices, which were based on much more expensively produced US oil. The result was low royalties for the producing countries and high profits for the Anglo-American majors.

When the system of payment to the Arabs and Iranians was changed from royalties to the so-called fifty-fifty deals, the producing countries became involved in the sale of their oil. Their revenues now depended as never before on how much their oil was sold for, and they did not like the system used for selling it by the subsidiary companies.

'You people just name your own price, then sell the oil to your parent companies, and we don't get much money,' they complained. Howard Page of Jersey Standard answered during one of the negotiations in which he was involved by saying:

'I'll tell you what we'll do. We will formally post a price for our oil in the Persian Gulf, and we will state that anyone who comes along can buy our crude at that price. That means there can't be any rigging of the prices and you will be able to see that you are getting a fair whack of the profits from the sales.'[1]

Since the formally and publicly posted prices from now on were high – although not as high as the more expensively produced US oil – the solution was accepted generally by Middle East governments, and for the first time many oil-producing countries were able to calculate how much their oil revenues were going to amount to in a given year and make up their budgets accordingly. Not everybody was satisfied that the posted price was high enough, and there were those who accused the parent companies, by a series of book-keeping tricks and transfers, of making hidden profits. But on the whole the posted price system was welcomed.

Until August 1960, that is, when Jersey Standard, without informing any of the governments concerned, knocked 14 cents off a barrel of oil – a drop of about $7\frac{1}{2}$ per cent a ton – as sold in the Persian Gulf or at the refinery terminal on the Mediterranean at Tripoli, there was a buyers' market abroad for oil, and surpluses everywhere. Russian oil was flooding Europe. The emergency situation caused by the Suez crisis was over. Libyan and Algerian oil were beginning to come on to the market. From an economic point of view – or, rather, from a strictly commercial standpoint – Jersey Standard had a good argument for the cuts. (They would have had a better one if they had not only cut the prices at which their subsidiaries were prepared to sell Middle East crude but also the price at which they sold the refined petroleum through their petrol pumps in Europe. But – and this should be remembered – this they did not do.)

It was suggested by some Arabs at the time that Standard's unilateral action was taken not strictly as a commercial price reduction but also as part of the all-industry pressure which was now being exerted by the Anglo-Americans on General Kassem of Iraq, to cut down Iraqi oil income and force him into line. If so, Jersey Standard does not appear to have informed its British partners of its decision, and BP, in fact, strongly protested against the reduction, before reluctantly dropping its own prices. Other companies followed suit.

If Jersey's action was meant to hit Kassem, however, the place where the most pained reaction came from was not Iraq but Saudi Arabia. It was a moment when King Saud was making one of his periodical attempts to reform. He had promised to balance the budget, end corruption, and bring about widespread improvements in education, health and communications. A programme had been drawn up – based on the estimated revenues the nation would receive from posted price sales – for vast building schemes and several costly but necessary projects.

Jersey Standard's price reduction slashed Saudi Arabia's income, at one stroke, by $30,000,000, for the year 1960–1. Loud was the lamentation. At once Sheikh Abdullah Tariki, who was now the Saudi minister of petroleum resources, rushed to Baghdad to announce his solidarity with Kassem in face of the pressure the Anglo-Americans were exerting on him. Other Middle East oil ministers followed him to the Iraqi capital. There they were joined by Dr Juan Perez Alonzo, minister of mines and hydrocarbons in the Venezuelan government. Venezuela was chief supplier of oil to the United States domestic market, and had a vested interest in persuading the Middle East to maintain high prices for its oil. Venezuela charged the US companies operating inside its territory a sixty per cent tax on their oil profits, and could not afford to be undercut by exports from Saudi Arabia, Kuwait, Iran or Iraq. He urged the Arabs and the Iranians to band together. But it was Tariki who was the leading spirit in the drive to form an official association.

From the emergency congress of Middle East oil ministers over which he presided in Baghdad in September 1960, emerged – squealing like a newborn babe – a body called the Organization of Petroleum Exporting Countries. Its member-ship consisted of Iraq, Iran, Kuwait, Saudi Arabia and Venezuela.* The members pledged themselves to demand stable prices from the oil companies and immediate restoration of the cuts. They insisted that no cuts should be made in future by the major oil companies without prior consultation with the governments of the producing countries. They acknowledged certain obligations to the companies (such as security of contracts) and demanded others in return: a steady income for themselves, a steady supply of oil in return to the consuming countries, and a 'fair return on their capital to those investing in the petroleum industry.' They pledged themselves to main-tain solidarity and remain united under all circumstances, and to spurn any special advantages any one among them might be offered by the petroleum cartel to persuade that individual country to break ranks.

It was the birth of OPEC. Not even the Arabs were confident of the baby's survival to begin with. Too many times had the peoples of the Middle East tried to band together in a united front, and been broken up by external bribes and internal rivalries and dissensions. How could this one expect to be different? As for the international oil companies, after a preliminary frisson of alarm, they refused to recognize OPEC's existence. They made it plain that they would continue to negotiate with individual countries and, in no circumstances, would they deal with a combined organization.

'We don't recognize this so-called OPEC,' said Bob Braun, at that time president of Aramco. 'Our dealings are with Saudi Arabia – not with outsiders.'[2]

Most of the oil majors decided that if they ignored OPEC it would break up and disappear, like other Arab organizations before it. They were all for taking no notice of it. Others con-

---

* Qatar joined shortly afterwards, followed by Abu Dhabi, Dubai, and (much later) Libya, Algeria and Indonesia.

sidered that the danger to be faced for the Anglo-American concessions came not from OPEC itself, but from the man who had been the leading figure in creating it, Abdullah Tariki. Word was passed around that if only Tariki could be shown up to the other Arab oil ministers as the anti-Western fanatic that he was, the whole revolt would collapse and that, at the very next opportunity, he should be discredited.

The opportunity came in October 1960, and never have the Anglo-Americans made a greater mistake than in seizing it.

On Thursday 20 October 1960, the Arab Petroleum Congress opened in Beirut to hear Sheikh Abdullah Tariki read a paper on 'The Pricing of Crude Oil and Refined Products'. Oil ministers from all the principal Middle East countries, heads of the major oil companies, and a flock of experts had been invited to listen to the paper and debate the points raised in subsequent discussions.[3]

Tariki had hoped that the meeting would be held under the auspices of the new association, OPEC, but the Anglo-Americans, eager though they were to get to grips with the Saudi minister, had made it clear that in such circumstances they would refuse to come, since their presence would imply a recognition of the organization. So the name, OPEC, though on everybody's lips in the corridors of the congress hall, was never mentioned in the speeches.

On both sides, the delegates were in a fighting mood. The cut in the posted prices by the oil companies had not only dealt a blow to the economies of the Arabs and the Iranians, but the way in which it had been done had been an affront to their pride. 'They didn't even inform us beforehand,' said one Arab minister, bitterly.* On the other hand, the oil company delegates, convinced that the Arabs did not understand the facts of commercial life, nor very concerned with Arab pride, were out to demonstrate that in a sink-or-swim world of competition between nations the governments of the Middle East had better learn how they could price themselves out of the markets.

Tariki's paper was just what they hoped it would be, long on rhetoric and short on facts. Copies of his polemical attack on the methods of the major oil companies had been distributed to the

---

* 'If we had told them beforehand,' said Howard Page of Jersey Standard, 'they would have tried to stop us from doing it.'

delegates beforehand, and the assembled British and American experts had prepared their replies with what they hoped was devastating thoroughness. But Tariki took some of the wind out of their sails by departing from his script in a sudden and dramatic aside in which he charged the oil companies with cheating by book-keeping. By arbitrary and discriminatory pricing of Middle East oil, he declared, and by transferring profits from oilwells to tankers, he claimed that the oil companies had made an extra profit over the past seven years of no less that $5,500,000,000, half of which should have been paid over to the Middle East producing countries, but had in fact been pocketed by the oil companies. The oil company experts were so flabbergasted by this fantastic attack – not even Rockefeller himself, after all, had ever been accused of misappropriating nearly three billion dollars – that when the day's session came to an end they insisted that an extra day should be given to the debate. After a night working on their documents and telephoning London and New York, they were convinced that Tariki would be shown up as a wildly dangerous windbag.

Next morning the big guns of the industry were wheeled in and opened fire on the Saudi minister. B. H. Groves of Socony Mobil read out a carefully prepared list of figures – which had been painstakingly picked up by telephone from New York during the course of the night – to prove that far from making a hidden profit of 5 billion through juggling its tanker figures, his company alone had made a $30,000,000 loss.

To the delight of the partisan, largely pro-Arab audience, Tariki was undismayed.

'What does it matter at what stage the profit – or the loss – is entered in your books?' he asked. 'You make a loss on your tanker transport operations in order to transfer the profits to your refining and marketing affiliates. The money was made somewhere, and you rarely sell to anybody who has no affiliation with your company.'[4]

Groves of Mobil gave way to George Ballou of Standard of California, who smoothly rebuked Tariki for using formulas for calculating oil prices which were never intended for that purpose, and assuring him that the oil companies were working in the best interests of the producing countries.

'How can we possibly take that for granted?' replied Tariki. 'See what happened in Saudi Arabia.'

He then proceeded to describe how, shortly after the conclusion of the fifty-fifty agreement between the Saudi government and Aramco, the posted price for Saudi crude was set at $1.75 a barrel. Then his department discovered that Aramco was putting down only $1.42 a barrel in their books.

'When we asked why,' went on Tariki, 'Aramco said this was a discount which they gave to the parent companies to build marketing facilities.' He paused and then went on: 'But that was never mentioned to us during the negotiations. And we didn't sign that in the agreement. But the effect was that what we were getting was not 50–50 but 32–68.'

He was beginning to warm up now.

'They tell us that they have to sell to the parent companies at $1.42 to build marketing facilities. But while they were doing this, they were telling the US government that $1.75 was the minimum price on which they could make a reasonable profit. You see? They can tell two governments two different things. Therefore we cannot accept anything they say without real investigation. They do not tell us what is going on outside the production phase. They say that this is complicated and is the business of the parent companies. They treat us like children. Now it's about time for the oil companies, if they really want the confidence of the people of the Middle East and their governments, to divorce themselves from the controlling end, and work for the governments and peoples of the countries in which they are operating.'

He sat down to thunderous applause, and leaned back to listen to a spokesman from BP (R. Anderson) assuring him that so far as his company was concerned, they were all partners together. Then he bounced to his feet again.

'You say we are in it together?' he asked. 'But when you make money you make it alone. When you lose, you come out and cut the posted prices. Somebody here mentioned that we don't need a formula. We do. We need a formula to protect ourselves, because if we don't you can easily sell our oil at fifty cents a barrel and take the profit through your tankers, refiners and marketers. You use the formula when you like it. You don't use it when we like it.'

237

The spokesman for Royal Dutch-Shell, W. Nuttall, rose and waited for the renewed applause to die down.

'May I say,' he began, 'that as far as the Shell Company is concerned, we have no intention of hiding things from him or anyone else. We would be delighted if Mr Tariki would come to our office in London, and we will be able to explain to him the points which he considers are being hidden.'

'Thank you very much,' replied Tariki. He waved his hand at the Aramco, Standard of New Jersey and Mobil representatives in the audience. 'The only thing I ask you to do is convince your friends here not to raise the posted price – not to raise it or lower it – before they have first come and consulted us – and convinced us that it is the thing to do.'

All the Arabs in the audience were on their feet now, applauding and cheering. The oil ministers crowded forward to wring Tariki's hand. Only the Western oil experts sat on their hands, shaking their heads in feigned astonishment at the enthusiasm all around them.

Not only had Tariki not been discredited. By shifting his ground with what one commentator afterwards called 'considerable skill', and never allowing himself 'to get caught in an indefensible position', he had neatly turned the tables on the Western oil experts. As the English correspondent remarked, he 'had brought into the open issues which have been festering for the past eighteen months or more ... saying, in effect; a) If you don't agree with our [Arab] figures, why not produce the correct figures relating to the profit and loss on each separate phase of the integrated operation; b) If you don't agree with our pricing formula, why don't you come and discuss the whole matter frankly with us, as the Japanese have done; and c) No matter whether you make a profit or loss on later operations, we want a share in them and want to know what is really going on.'[5]

The basic instability of the relationships between the host countries and the Anglo-American oil companies which operate on their territories had been laid bare.

'The chips are down,' wrote Ian Seymour, 'and what happens next is anybody's guess.'[6]

Abdullah Tariki was now the hero of the hour throughout the Middle East, and the association which he had done most to create, OPEC, flourished in his reflected glory. The governments of the oil-producing countries, which had reluctantly accepted the idea but were sceptical of its survival, suddenly began to share the enthusiasm of their oil ministers. It is an indication of their new confidence that the shah and the sheikhs even consented to advance money to the new organization. As a result, OPEC was able to vote itself a budget of £150,000 and establish its headquarters and secretariat in Geneva, Switzerland.*

It only remained to choose a secretary-general for the organization. The logical choice would have seemed to be the man who brought it into being in the first place, Sheikh Abdullah Tariki. But while the members were debating this point, the fiery Saudi was back home, savouring his triumph, and, emboldened by the plaudits he had received, prepared to launch himself upon a crusade against his favourite enemy, Aramco. Each day he received Arab journalists from Baghdad, Mecca, Cairo and Beirut, and to each of them he poured out complaints against the American company.

The Saudi people were being cheated, he protested. Take the Transarabian Pipeline (Tapline) from Dhahran to the Mediterranean, for instance. Tapline was owned by Aramco, but did the company share its profits with Saudi Arabia on the oil it lifted from the Saudi fields? It did not. It passed them on to the parent companies in the United States.

'This is unfair practice,' he declared. 'We demand that Tapline's profits be split fifty-fifty with us. The company already owes us $180,000,000 which is our share of the gains they have made in the past few years.'⁷

He asked that the money be paid over at once. He followed this by demanding a complete revision of Aramco's concession agreement, the relinquishing of territory, the stepping up of payments from 50–50 to 75–25 and even higher percentages, and a swifter 'Arabization' of the company.

Thoroughly alarmed at all the publicity these demands were getting, Aramco issued a statement setting forth the enormous efforts it had made in recent years to 'Arabize' its operations, and detailing its welfare work on behalf of Saudi Arabia

---

* When Switzerland hesitated to give blanket diplomatic privileges to the secretariat, OPEC moved to Austria (in June 1965), where the government proved to be more amenable.

generally. It appealed for calm discussions of any differences which might exist between the company and the government, 'in a spirit of cooperation and goodwill'. The statement continued:

On this issue the government and Aramco hold honest differences of opinion, since Aramco believes it owes the government nothing in this connection. In view of this, Aramco wishes to assure the gracious Saudi Arab people that it stands ready to discuss or arbitrate, if necessary, any point of difference with the Saudi Arabian government.[8]

Tariki replied to this by hotly denying Aramco any right to arbitrate on tax matters which were the sole concern of the Saudi government, and threatened to report Aramco's 'iniquitous exploitation' to OPEC.

But suddenly there was anxiety in court circles in Riyadh and Mecca. Funds were short again. Once more King Saud had allowed his corrupt friends to gain the upper hand, and his budget was over-balanced. He needed allies. He needed discipline in his government. So on 15 March 1962, he came to terms with his younger brother, Crown Prince Faisal, and announced a new administration with himself as prime minister and his brother as deputy prime minister. Abdullah Tariki was no favourite of Prince Faisal, and when the names of the new ministers were announced, Tariki's name was not among them. No comment was allowed in the Saudi press about his dismissal. He quietly left Saudi Arabia shortly afterwards to go into exile, and set up in Beirut, Lebanon, as an oil consultant. Aramco was not the only Western oil company to breathe a sigh of relief at his disappearance from the limelight.*

In the meantime, OPEC had gone elsewhere to find its new secretary-general. The members chose Fouad Rouhani of Iran, and made him both secretary and chairman of the organization. He was a very different character from the explosive Tariki, a quiet, urbane, cultivated man with a taste for Western culture, wines, food and literature.

It would be Rouhani's job to persuade the Anglo-Americans to recognize the new organization, instead, as Tariki had hoped, of forcing them to sit down as supplicants at OPEC's table.

---

* Tariki, who now advises the Algerian and Iraqi governments on oil matters, was in the winter of 1971 forbidden to reside in the Lebanon, it is believed at the request of the Saudi Arabian government.

PART SIX

The Challenge of the Sixties

2 1 · Death of a Gadfly

If anyone could be blamed for precipitating the cuts in the posted prices of Middle East oil in 1960 – and thus for the Arab attacks on the major petroleum companies which stemmed from them – it was Enrico Mattei of Italy. The Italian oil chief was still the implacable enemy of the big Anglo-American combines, and his determination to make life difficult for them whenever the opportunity offered had now become a crusade. He buzzed angrily over the operations of the big companies like a gadfly over a cud-chewing cow, swooping down to sting whenever the animal looked too self-satisfied. In truth, the viciousness of his attacks on the Anglo-Americans was nowadays not simply a reflection of how much he resented them, but also of his chagrin at the lack of success of his own enterprises in the world of oil.

Though it was Mattei who had pioneered the 'joint venture' with Iran which had changed the nature of Middle East petroleum exploitation, neither he nor Italy had benefited to any great extent from the partnership. No money had been lost on the outlay, certainly. But oil discoveries in SIRIP's concession area had been modest rather than bountiful, and he had been forced to seek new pastures – in Egypt, Libya, the Sudan and Morocco – where he hoped to reap the rich supplies of oil which he needed to keep ENI, the Italian national oil company, of which he was still president, and Italy itself supplied with fuel. So far the bonanza had not been forthcoming, and it did not lessen Mattei's antipathy towards them that the major oil companies, despite difficulties, continued to thrive. He missed no opportunity to do them damage, and in the late 1950s he hit them with a very low blow indeed. He went into the market places where the big companies sold their oil and petrol, and he undercut them. The chief target of his operations was

Standard of New Jersey, operating in Europe under the name of Esso International, and if ever there was a case of a biter being bit, this was it. Jersey was a chip off an old American oil bloc (the original Standard Oil) which had won its way to international power by using its vast resources to undersell and bankrupt its competitors.

Mattei did not have the vast reserves that John D. Rockefeller had once used to destroy the rivals of Standard Oil, and Standard of New Jersey was never in any danger of going bankrupt as a result of his operations. But he was, nevertheless, able to make commercial life for Jersey uncomfortable in the extreme. How could he do so, since ENI had so far failed to find a richly productive oilfield of its own?

The answer was Russia. It was a period when the Soviet Union was searching avidly for a market outside its European satellites for its surplus supplies of crude oil, of which it possessed enormous quantities. It needed foreign currency. Enrico Mattei flew to Moscow and had a series of talks with the Soviet leader, Kruschev, and it was a meeting of minds, or rather of mutual antipathies. From both their points of view, it would be a productive relationship. Russia had the oil, which it was prepared to sell cheaply. Mattei had the outlets – and would be buying at bargain prices. Both of them would be earning the extra dividend of damaging the Americans.

In 1959, 1960 and 1961 the crude began arriving in Italy from Russia in ever increasing loads. It was refined in ENI's refineries, and then, as a result of a series of deals made all over Europe by Mattei, flooded on to the market at prices well below those maintained by the major oil combines. Particularly hard hit by the undercutting were Standard of Jersey and IPC, both of which lifted vast quantities of crude to the Mediterranean from the Gulf and Iraq especially for the European market. In the days of John D. Rockefeller there would immediately have ensued a savage price war in which – backed by unlimited supplies of oil – Standard Oil would have flooded the market with cheap petrol until the enemy ran out of petrol and money.

But Mattei had access to all the crude he needed, and the Soviet Union, in the circumstances, was not worried about how cheaply it was sold. On the other hand, Standard of New Jersey (as a member of Aramco and other concessionaire groups in the Middle East) had the posted price system to maintain and knew it could not reduce the price of its crude without reducing its own profits and cutting into the oil revenues of the Middle East producing countries. It was a dilemma facing all the major companies to a larger or lesser degree, but for Standard of New Jersey it was by far the most economically damaging. Finally, as the flood of cheap Italo-Soviet oil rose and spilled into Esso's markets, the American company decided to take the risk. It cut the posted price of Gulf-produced crude first by ten and then by seven-and-a-half per cent. It was the worst thing it could have done if the oil companies wished to retain the status quo of their concessions in the Middle East. It brought the Arabs and the Iranians up in fury against them as never before, and united them in OPEC.

Nor did it do any good against Mattei. With Russian connivance, the Italian cut the price of his refined crude to meet Esso's challenge. By now, it was obvious even to Standard of New Jersey that the company dare not follow suit and slice its prices again, and again, and again, until the Italian admitted defeat. The Middle East governments would never stand for any more cuts. They had made that plain at Beirut, and ever since. In any future change of price, they would first have to be consulted, and so far as cuts were concerned, they would never give their permission.

All through the bleak commercial months of 1961–2 Jersey brooded and pondered over what to do about Mattei's 'unfair competition', and hard were the words that some of the executives used about him. 'Moscow's stooge' and 'tool of international Communism' were two of the politest pejoratives. Finally, cursing the stupid blunder which had made an enemy of him in the first place, the executive committee of Standard of New Jersey decided that there was only one thing to do: if they couldn't beat Mattei, they must persuade him to join them. But how? He was touchier than an Arab oil sheikh. He never forgave snubs, and Jersey had snubbed him three times in the course of the years – and two of the snubs had been against him personally. In the between-the-war years, when Mattei was an oil official under Benito Mussolini, Jersey officials had discovered that he was standing in the way of a project they had conceived for obtaining a blanket concession for the oil resources of the Po valley, in northern Italy, and some hard words had been said by an American executive about his antecedents. Mattei, son of a peasant, half-proud, half-ashamed of his humble origins, had taken the words seriously and had not forgotten or forgiven. Years later, as head of Italy's ENI, he had gone to New York and visited the headquarters of Jersey, where he expected to be received as the important personage he had become. But ENI to Jersey was just another small oil company, and he had been kept waiting. That he did not forget either.

In the circumstances, it was no moment to expect that all would be forgiven over a handshake. Some much more subtle approach would have to be made. But it would have to be made soon, before the flood of Mattei's cheap crude played havoc with the markets.

It so happened that the major oil companies were not the only ones who feared Enrico Mattei. General de Gaulle and the French had even more valid reasons for objecting to his methods. These were the days, in the early 1960s, when Algeria was still French but the fight for its independence was reaching a bloody climax. In the Algerian Sahara the French had begun intensive exploitation of the desert oilfields, and no matter what happened to the rest of the country, these they were determined to hold on to. To give an international flavour to the operations, an emissary of General de Gaulle had been to Rome, where he offered Mattei and ENI a concession in the desert, in partnership with the French.

Mattei had turned it down in no uncertain terms. He told the French two things which infuriated them, because they were eventualities they feared but were determined they would never allow to happen: a) that they would lose the war in Algeria, and b) that they would then lose the oilfields.*
When news of his remarks leaked out in Paris and Algiers, the French government officials preserved a tight-lipped silence, but members of the army, particularly the 'ultras' of the OAS, the right-wing military terrorist organization, were loud in their denunciations of him. Soon, when rumours spread that Mattei was in touch with Algerian rebels, the OAS began to make threats against his life. They were threats not to be taken lightly, since the 'ultras' had already shown their ruthlessness

---

* He was quite right in both cases, of course.

in several particularly nasty kidnappings and assassinations in France, Switzerland and Belgium.

Mattei was unperturbed. In August 1961 he gave an interview to Gilles Martinet of the Paris weekly, *France Observateur*, during the course of which these passages occurred:

INTERVIEWER: There is at least one point on which General de Gaulle's ministers and those we call the 'ultras' find themselves speaking in identical terms—namely when they talk about you. In their eyes you are the man who plans to destroy France's interests in the Arab world and who, more specifically, has his eye on the immense resources of the Sahara.

MATTEI: What I am really being blamed for is my refusal to commit myself to a certain policy, in other words for having refused to set myself up in the Sahara along with the French, English and American companies. The offer was made to me several times, but I consistently declined to accept a concession. I don't want my technicians to find themselves faced with the necessity of working under the protection of machine guns. Italy lost her colonies together with the war. Some people considered this a misfortune; in fact it turned out to be an immense advantage. It is because we no longer have any colonies that we are so warmly welcomed in Iran, the UAR, Tunisia, Morocco, Ghana and other countries. I don't see why we should endanger this position by joining in an operation which everyone knows cannot be continued indefinitely in its present form.

When the Algerian war comes to an end, I shall see what can be done.[1]

Shortly after this interview was published, Mattei flew to Tunis, where ENI had oil interests and concessions, but where there were also several scores of thousands of Algerian rebels training in camps for war against the French. It seems most likely that Mattei met some of the Algerian leaders during this visit, and that he had further talks with rebel representatives in various parts of Europe during the next twelve months. He was reported to be on the OAS death-list, but was unworried about it, though he now moved everywhere with a bodyguard.

It was against this background, as well as the more mundane one of cheap Russian oil, that Standard of New Jersey set about wooing Enrico Mattei over to the American side in the summer of 1962. The depredations of Soviet crude under Mattei's control were now such that the Oil Lobby had been in operation behind the scenes in Washington, and the aid of President John F. Kennedy had been secured. Some bright Jersey executive, remembering Mattei's secret shame over his humble beginnings, and the poverty which had prevented him from going to a university, had come up with a brilliant idea. A plan had been devised to invite him to the United States to receive an ad honorem degree from Stanford University, which has always been interested in petroleum affairs; and afterwards he would go on to Washington, where he would be received by the president. From there on, the oil company executives would take over.

In furtherance of the plan, Kennedy sent his under-secretary at the State Department, George Ball, to Rome, where he dined in secret with Mattei at the US embassy.[2] A few days later, the Italian consented to meet Standard of New Jersey officials. It seemed that flattery had got them everywhere. Soon Mattei was confiding to his friends: 'The Americans are now offering to sell me crude at a price which matches Russia's price.'

Of subsequent talks, a document in the archives of the Middle East Economic Survey reveals that an agreement with Esso was being discussed.

'[It] was to cover such questions as long-term supplies of crude oil to ENI (probably from Libya), the purchase from the engineering company of ENI group Nuova Pignone of vast quantities of petroleum-industry equipment, and the supply to the ENI marketing company, AGIP, of refined products in areas in which ENI still had no refinery.'[3]

According to a statement made later by Standard of New Jersey, the agreement was to be signed at the beginning of November, when Enrico Mattei would go to the US for his honorary degree. But on 27 October 1962, he was killed. He took off from an airfield in Sicily, where he had been visiting ENI installations, in a company plane. It crashed shortly afterwards, killing all aboard. And no one has discovered since whether it was an act of God or of sabotage.

Standard of New Jersey was quick to issue a statement (through a spokesman of its Italian subsidiary, Standard Italiana, on 29 October) deeply regretting his untimely death, and emphasizing that the state of belligerency between the company and Mattei was on the brink of changing to the closest

I

collaboration. But a milch-cow does not genuinely mourn the swatting of a gadfly which has been buzzing around her for years, and there were not many long faces in the front offices of the big oil companies when news of the Italian's death came through. Executives of other companies read Standard's statement with a certain scepticism, for they doubted that any such agreement could have penned up Mattei for long. He had spent most of his life fighting the Anglo-American oil bloc, and he had never made any secret of his feelings towards it.

In concentrating in a few hands the control of oil production and marketing, [he had once said] in maintaining with the consumers the relations of supplier and customer in a closed and rigid market, in granting only financial returns to the countries that own the oil, and in barring all international agreements for rational organization of the market, the international companies have increased their own power but they have also created the conditions for either the break-up or the transformation of the system under the pressure of new forces.[4]

He had always considered himself one of the 'new forces' and no agreement with a member of the international cartel would have been likely to relieve the pressure on them for a moment longer than Mattei found it expedient for his purposes. His death was a tragedy, but it would be hypocritical to pretend that the major oil companies were anything but relieved by the news that he was dead.

22 · Overthrow

So far as their public declarations were concerned, the men who ran the Arabian American Oil Company (Aramco) considered King Saud of Saudi Arabia a forward-looking and enlightened monarch only interested in the welfare of his country and his people. Aramco's president, F. A. Davies, had used words of the highest praise in assessing Saud's qualities for the benefit of a Senatorial Committee in Washington (see p. 157). But everyone in the company knew that the king was, in fact, a disastrous failure whose regime was riddled with corruption and racked with scandals that were the talk of the Middle East. Though the oil revenues from Aramco and from the independents operating in the Neutral Zone were still, as in his father's day, paid to him personally – and were running now at a rate of $350,000,000 a year – he had managed to get both himself and his country hopelessly into debt. By 1960 Saudi Arabia owed 1,250,000,000 riyals abroad and over 500,000,000 riyals at home, and so much paper money was being printed that the value of the riyal had fallen from 3.75 to 6.25 to the dollar.

Yet Saud did not for one moment think of curtailing his wild expenditure, on cars, planes, wives and concubines. He now had committees in Cairo, Beirut, Teheran and Karachi buying up girls for his harem, and the export of young females (and small boys) from Cairo to satisfy the demands of the king's five thousand spendthrift courtiers had now become so notorious that the Egyptian press was demanding its suppression.

Aramco was not particularly worried about the king's morals,* but his greed for money, and his willingness to agree

---

* Though they suffered badly from his hypocrisy. While indulging in the most extravagant excesses himself, he imposed a puritanical regime on oil company employees, particularly the Americans and other non-Saudis, forbidding wives to drive cars, Christians to have a church, and banning all alcoholic drinks, including beer.

249

to any scheme in order to get it, caused them considerable concern. Company officials had had to work hard to sabotage a scheme by which Saud planned to give the Greek shipowner, Aristotle Onassis, a monopoly for the use of tankers to transport all oil out of Saudi Arabian territory. The scheme had, in fact, been announced in the last days of his father, Ibn Saud, but it was his son and his cronies at the Saudi court who had arranged it, and had fixed up the details with Onassis's go-betweens. They included the late Hjalmar Schacht, once financial adviser to Adolf Hitler, and a Greek named Spyridon Catapodis (who later sued Onassis for fees alleged to have been promised amounting to $14,210,000). And it was Saud who took most of the bribes which are said to have been passed over to secure the tanker contract, amounting to several millions of dollars. Aramco's reaction to the grim possibility that Onassis might soon be in a position to decide in which ships and at what rates they could export their oil was to threaten to take King Saud and his government before an international court. At the same time Aramco's parent companies started to squeeze Onassis before he could squeeze them by instituting a quiet but effective boycott of the Greek's tankers, and it was this rather than the threat of legal action which finally broke up the deal.*

It was not only a costly episode for Aramco but King Saud's lies and deceptions during the course of the negotiations filled the directors with the direst fears for the future of their operation under the aegis of such an untrustworthy monarch. He was likely to cause trouble in the future not only for the oil company but for the country, for the way in which he flaunted his money and his overweening vanity were making him notorious in the Arab world. He was reputed to be jealous of the influence of Gamel Abdul Nasser of Egypt, feeling that he and not Nasser should be regarded as the leader of the Arab world, and he had allowed himself to be involved in an inept plot against the Egyptian's life. As a result, powerful organs of the Cairo press and radio were now pouring out attacks against him.

* Onassis was afterwards to describe it as the most critical moment of his career. 'Because I was in the doghouse with the oil companies and could get no new charters, I had a great number of idle ships on my hands,' he told his biographer, Willi Frischauer.

Gangs have been set up here to export white slaves to him—this in the day and age when man's exploits have taken him to the reaches of outer space [wrote Al Ahram]. Arab oil has given rise to many complexes. Those who acquire new wealth develop an intense loathing for the parasites squirming at their feet, waiting for the handouts. And those who stretch out their predatory hands—the opportunists—succumb to greed and to the lure of fast money, abandon all sense of honour, and, above all, develop a contempt for the oil kings whose coffers are overloaded with gold through no effort of their own.[1]

What Aramco feared most was the unrest that this barrage of anti-royalist propaganda might cause in Saudi Arabia itself, for Aramco was now so closely linked with the monarchical system that any revolt against the regime would almost certainly turn out to be a revolt against the oil company as well. 'Every time the king sneezes, Aramco builds another hospital,' one British critic sniffed.

But none of the Americans dared raise his voice against the royal excesses. The one non-Saudi who had had the courage to do so was Harry St John Philby, but despite the fact that he had been the trusted adviser of the old king, his son and heir served Philby with an order of expulsion and vilified him as a liar, a traitor and a cheat. The scandalized and angry old Anglo-Irishman went grumbling off to the Lebanon, mourning the sad depths to which his adopted and beloved Saudi Arabia had fallen.*

Fortunately, there was one man in the land whom King Saud both envied and feared, and that was his younger brother, the Crown Prince Faisal. The two men shared only one family characteristic, a tendency to sickliness; otherwise they were the opposite in appearance, in temperament, in education and in outlook. Beside his fleshy, flaccid brother, Prince Faisal had the lean and hungry look of a famished eagle, and indeed a desperate need to appease some ravenous worm inside him had always been one of his troubles.

'A handful of rice goes into his mouth like a cannonball,' his father once said of him. 'He is a hero of the spread. He eats – he eats!'[2]

Otherwise, he was rarely given to excess. In 1960 he had

---

* When Philby proved to be even more vocally critical in exile, King Saud invited him to return, on the theory that he was more easily controllable inside the country.

eight sons compared with Saud's forty or more, and he lived happily with one wife as against Saud's ever-changing wives, concubines, slaves and cabaret girls. He was one of the few sons whom the old king had encouraged to travel abroad, and he had a social and political horizon beyond the minarets of Mecca and the sand dunes of the Empty Quarter. He lived modestly in a small residence which seemed humble compared with the king's flashy palaces, and though he shared the Saudi family's love for gold, he rarely if ever flashed around his considerable wealth.

There was never any doubt in either brother's mind that Faisal would have made the better king, and the knowledge of it sometimes embittered relations between the two men. But each time King Saud faced a crisis, he called Faisal in to help him, and a regime of austerity would follow, with the crown prince established as prime minister or deputy minister, empowered to cut the allowances of the vast numbers of feckless princes, and even to limit the king's own privy purse to a maximum of $60,000,000 a year. Then, when the crisis was passed, and everybody had begun to complain of the miserliness of the regime, he would be banished into obscurity and the orgy of spending would begin again.

It was not that Prince Faisal was an enlightened or a liberal element in Saudi Arabian life. He disapproved of his brother's profligacy not because he wished to replace it with a more democratic regime; and he certainly had no plan for allowing the oil revenues of the country to pass out of royal hands. He had listened to the polemics of the erstwhile Saudi oil minister, Abdullah Tariki, with prim distaste, and he disliked the minister not simply on personal grounds but for political reasons as well. He was convinced that Tariki was a secret spokesman for the foreign-trained Saudis, most of them educated abroad, who had started an underground National Liberation Front, and he had tolerated him only so long as he was forcing the oil companies to pay higher prices for their oil and he had considered him a danger to the regime as well as to Aramco when he talked of takeovers. Faisal was a tidy, efficient, money-minded conservative who abhorred the waste of his brother's regime, but wanted no change in its basic structure.

In the spring of 1962, the king appointed himself prime

minister, and Prince Faisal accepted the post of his deputy, but only on condition that Tariki was dismissed and banished from the country. Faisal knew that it would only be a matter of time now before he was in charge of the country, for King Saud was neither physically nor mentally equipped to cope with the tasks of the premiership. The king was recently back from a visit to the United States, where he had undergone operations for trachoma, that endemic Arabic infection, at the Peter Brent Brigham Hospital in Boston, and it was as much as he could do to shuffle from the festive board to the bed of his latest favourite, let alone having to handle urgent affairs of state. They were urgent because, at Faisal's insistence, a programme of school and hospital building had begun, and yet another austerity programme would be necessary to pay for the projects. At the same time, Egypt was backing the rebels in the neighbouring state of Yemen, while Saudi Arabia was discreetly but expensively subsidizing the Yemeni royalists. Saud, still smarting from Nasser's attacks upon him, was all for sending Saudi forces across the desert frontier to confront the Egyptian army, and had petulant outbursts of temper when Faisal succeeded in circumventing him.

By 1964, the tantrums had become too much even for the patient Faisal to bear. He sensed that one of those moments was approaching when the king would once more dismiss his brother, and revert to his feckless ways. But this time, Faisal decided, the times were too fraught, and Saud too irresponsible, to allow it to happen. In the last days of October, he sent members of his entourage around the salons and coffee houses of Riyadh, Mecca and Jiddah, sounding out who were his enemies and how much he could count on his friends. In the past months he had bound a number of the more influential princes to him by quietly restoring their allowances and promising them posts in any new regime which might be established should 'our beloved brother fall victim to his increasingly dolorous afflictions'. He had also talked with the Muslim priests, the ulema, assuring them that should he ever be ruler of Saudi Arabia, the state would more than ever be subject to the laws of the Koran and the strict rules of the Wahhabi sect.

On the night of 2 November 1964, a meeting of princes and ulema was held and it was unanimously voted to depose the king and acclaim Faisal monarch in his place. Shortly afterwards King Saud's palace was surrounded by selected troops, and Prince Faisal, conceiving it his duty, personally presented an ultimatum to his brother. Either he abdicate and consent to leave the country, or he would never leave his palace alive.

King Saud capitulated and signed the instrument of abdication at once. He did not reproach his brother, and made only one request of him – that he be allowed to take certain members of his harem (of which he would prepare a list) into exile with him. They and some newcomers would help him spend the $3,000,000 a year which Faisal gave him. But not for long. He died in Greece in February 1969.

Spurred on by government propaganda organs, the Saudi people hailed their new monarch, King Faisal I. In Dhahran, the American oilmen of Aramco hailed him too. They knew that his regime would be theocratic, puritanical, and there was less likelihood than ever that they would be allowed to drink an occasional glass of beer. But at least the new regime would be a tidy one, and the days of waste and corruption – they hoped – were over.

They went back to the business of pumping out oil, and gave the order to step up production. Aramco was still looking for new fields, but for the moment it had all the crude it could handle.

In that, Aramco was luckier than some rival Americans working across the desert to the southeast, in the mountains of Dhofar.

It is one thing to be given an oil concession for a province the size of the state of Ohio, but quite another to turn it into a viable oilfield. As Dr Wendell Phillips and his friend, Sultan Said bin Taimur of Oman, were discovering, oil is a funny thing. Developments in geophysics, the use of magnetic surveys, gravity surveys, seismic surveys, and all the appliances of modern science, make it possible to plot from the air the structural traps in which oil may have accumulated. Geologists on the ground can judge from rock samples and structural prospects whether the traps look promising. But only by drilling wild-cat wells can it be discovered for certain that an accumulation of oil is definitely there, and even then the amount may be small, the quality may be poor. The more wildcats drilled in any area the more chance there is of finding rich accumulations. But the cost of them is enormously expensive, in men, materials, in time and effort.

This was what Dr Wendell Phillips and his associates were discovering in the province of Dhofar, and by 1961 the Dhofar-Cities Service Petroleum Corporation* had spent $30,000,000 on wild-cat wells without finding a viable field. They were beginning to run out of money and the sultan out of patience.

At first it had seemed to go so well, and the sultan had shown himself at his most urbane in his efforts to make life easy for the influx of oilmen who came flooding into Dhofar. This part of the Arabian peninsula, at least along its coastal strip, is a paradise compared with the bleached and baking desert of the Empty Quarter, which lies behind it to the north-west. The sultan's castle stood on the seashore outside the port of Salala, and was hidden away behind high walls at the end of a palm-lined avenue; and here, in a crenellated building looking down on the Indian Ocean, cooled by the tail end of the monsoon, Phillips would foregather with his bearded friend to report progress. His camp had been established in a small rocky cove some ten miles up the coast in Rihut, at the foot of a cliff from which jutted the remains of an old Portuguese fort. From it the crews had built a road up through the Qara mountains, a range four thousand feet high, to the arid desert plain beyond, and there they had drilled, above a promising Marmul structure from which the geologists were confident oil would flow. As indeed it did. From the first two wells that went down came crude in abundance, and the sultan had been summoned to see the liquid flow and he had rejoiced. He had sent greetings to all his fellow Arab rulers, messages to Kuwait, Dubai, Abu Dhabi, Riyadh and the Yemen, proudly announcing the discovery and informing them all that henceforth he would be a member of that exclusive club to which only the oil sheikhs belonged.

Unfortunately for everybody, the oil which flowed so freely was not the right kind of oil. Had it been discovered in the United States, or close to a great centre of civilization, it would have made its exploiters rich. It was low quality crude, marvellous for building roads but of no value whatsoever as fuel, and not worth building a pipeline to take it to the coast or a port from which to ship it to the outside world.

* The name of the group which had taken over the Phillips concession.

By late 1961, Cities Service had brought Richfield in to join the hunt for better quality crude, and they had drilled twenty-three wells, six of them over ten thousand feet deep, with no important results. To make things more difficult, a local imam, with rival claims to Dhofar, had begun a rebellion against the sultan, and the drilling crews now worked under armed guards, against tribal marksmen hidden in the mountains around them. Phillips, a tall, pale, frail-looking man, wandered around the camp with his two six-shooters on his hip, potting at stones, or teaching the sultan's son, Qaboos, lessons in marksmanship. For one period, Phillips was overshadowed, and the sultan's anxieties were momentarily appeased, by the arrival of a burnished blonde and a cheerful older woman, who turned out to be Phillips' sister and mother. So long as they were on the scene Sultan Said bin Taimur kept his temper. But when they left, he disappeared into his cool palace and allowed his officials to plague the company with petty restrictions. The group had now spent $40,000,000, and Phillips was beginning to feel guilty about the million dollars in his account at Chase Manhattan.

In 1962, he was overshadowed once more, this time by a larger-than-life Texan named John W. Mecom. Mecom was the man who had once scornfully remarked about J. Paul Getty that 'he doesn't know where his next half-billion is coming from.' He described himself as 'a modest billionaire', but in fact rarely hid his light under a bushel, and indeed had no need to. The outline of his life would make a classic script for a John Wayne film. He had started life as a roustabout in the Texas oilfields and had progressed to wildcatting in south Texas and south Louisiana, where he had drilled the world's deepest well, down to 25,000 feet. Since he was a one-man corporation who (like Getty and Gulbenkian) carried his office around with him, he did not have to worry about other people and risked his own money rather than someone else's. It was his experience that of every twenty wells you drilled, all but one would be dry or full of salt water or otherwise unproductive, but the one successful well would make you so much money that it would more than make up the expense of the others within six months of going into production, and would be 'pumping money' from then onwards.

'Oil is a hell of an expensive quantity to find and exploit,' he once said, 'but once you've found it you have gained possession of the philosopher's stone.'

Mecom looked over the Dhofar concession and proposed to Phillips and his associates that he should drill a separate well (in association with the Pure Oil Company of Texas) on a site of his own choosing, and some months later he took over the whole operation from Cities Service. He drilled five wells in the next three years, and found promising showings of crude in three places. But no bonanza. By 1965, even Mecom was discouraged, Wendell Phillips was hugging his million dollars for comfort, and Sultan Said bin Taimur was crushed.

It had all cost something around $50,000,000, and so far as the lives of the unfortunate subjects of Muscat and Oman were concerned, it had not altered a thing. But then, only the departure of the sultan would do that.

23 · The Libyan Jackpot

All the important Arab oil countries and Iran – plus a non-Middle East producer, Venezuela – now belonged to the Organization of Petroleum Exporting Countries (OPEC), and, as the oil companies were discovering to their cost, their combined weight was beginning to be considerably greater than the sum of its parts. With every month, OPEC's experience and know-how grew, and it was learning, slowly and painfully, and not without back-sliding by one country or another, tempted by Anglo-American blandishments, that in union there was strength. In the years since its formation in 1960, OPEC's progress had been impressive. There was still no formal recognition of its existence by the major oil companies, and in their negotiations with the various governments of the Middle East the companies insisted that any agreements were limited to the frontiers of the countries in which they were made, and they refused to contemplate global agreements.

To strengthen OPEC's position, the new oil minister of Saudi Arabia decided to resort to a stratagem. Ahmad Zaki Yamani was a Harvard-trained lawyer, a specialist in international law, who had taken over after the dismissal of Abdullah Tariki by the new King Faisal I. He was (and still is) one of the coolest and most persuasive Arabs in the Middle East, totally unpolemical, anxious to demonstrate the basic reasonableness of his attitude, a handsome and soft-spoken negotiator of the utmost charm and guile. It was Yamani who had helped OPEC surmount its first major obstacle in 1962, when the association passed a resolution designed to force the Anglo-Americans into a radical new system of 'expensing' their oil royalties. It was a move that would considerably increase the incomes of the producing countries, and OPEC wanted to negotiate the new system as a body, both to get the organization accepted and in

258

order that smaller or weaker member countries would not be tempted to settle with the companies for less than the others.

Aramco was chosen as the first company with whom we would reach an agreement, [said Yamani later] and I informed the company that an OPEC committee would like to have a meeting. I think Aramco might have accepted if it had been left to them, but they informed their member companies, and it scared them out of their wits. So Bob Braun, the Aramco president, refused to have the name of OPEC even mentioned in the negotiations. He said, 'We don't recognize the so-called OPEC,' rather in the same stupid way that we Arabs nowadays say, 'We don't recognize the so-called Israel.'[1]

Yamini went ahead anyway and formed a three-man committee on behalf of Saudi Arabia and the other member countries, and then marched them into the conference room.

The Aramco people at once challenged the legality of our three-man committee. But I was the chairman and I said I could choose whom I wanted, and we were not going to begin talks until they had accepted us. I told everyone to sit tight. We forced it on them. I did it just to impose OPEC on them. And by the end we had de facto if not yet de jure recognition of the organization.[2]

By 1962 OPEC members between them were the landlords of ninety per cent of the world's exported oil. (The United States was not, and is not, a major oil exporter.) However, there was one country which was beginning to emerge about this time as one of the most important oil exporting countries in the world, and that was the North African state of Libya. But Libya was the Achilles' heel of OPEC, because it was not a member of the organization, and had refused repeated invitations to join. In the meantime, it was selling crude oil well below the posted price of the OPEC countries, and thus threatening to sabotage every agreement that organization made.

The truth was that Libya, through a mixture of economic ignorance, naivety, and the venality of its administrators, had been sold short over its oil resources. It would be too harsh to say that the big American companies who were now established in the deserts of Tripolitania and Cyrenaica, the two Libyan provinces, had cheated the Libyans, but they had certainly achieved some extremely sharp bargains.

Until it was liberated by the Allied armies in 1943, Libya had

been an Italian colony in which the native Senussi tribesmen
had been degraded into worse than second-class citizens, and
all the more fertile areas of the country – particularly the Jabal
Akhdar, or Green Mountains, in Cyrenaica – turned over to
surplus Italian peasants for model farming. The Libyans, on
Mussolini's orders, had been denied education or any say in the
administration of their land, and their spiritual leader, King
Idriss, driven into exile. At the end of World War II, King Idriss
had ridden triumphant back into his capital, Tripoli, on the
same white horse upon which the Italian dictator had once
boasted he would lead a victory parade through Cairo. Libya
was free, and Libya was independent. But what can you do
with freedom and independence when you have only a handful
of administrators capable of running a country, and no schools,
no technical colleges, no technicians to run even the smallest
industries? Libya had no alternative but to call in European –
principally British – experts to advise them. Italian craftsmen
and mechanics were allowed to stay on, even though they had
once been hated oppressors, to keep the electric lights burning
and the cars and trucks running. In return for their aid, the
British were given the right to station troops in the places
where they had fought some of the most spectacular battles of
the Western Desert, at Tobruk and Benghazi, and here they
also had a base for the Royal Air Force and a staging post for
trans-African air services. The US Army Air Force, as part of
the so-called Eisenhower doctrine of Middle East peace-keep-
ing, had a big base at Wheelus, near Tripoli. King Idriss ruled,
but in little more than name.

Then, in the mid-1950s, oil was discovered in Cyrenaica. As
in the biblical lands of Arabia, there had always been gas fires
and bitumen seepages in the Western Desert, and it seemed
certain that beneath them were sizable accumulations of oil.
From some of the samples which had been tested, there was
every reason to believe that the quality of the crude, when
once brought to the surface, would be of a high grade, with a
low sulphur content, and therefore most suitable for sale to
the European market.

When news of the possibilities spread through the oil world,
in came the companies in search of concessions. But Libya was
uncertain as to how to go about granting them. The king's
ministers were told by their foreign advisers – none of them,

unfortunately, oil experts themeslves – that they should at once set about writing a general petroleum law, defining the terms under which the concessions would be granted. The Libyans were by this time extremely anxious to see their potential petroleum wealth turned into hard cash, and they were therefore most grateful when the American oil companies offered to speed things up for them by loaning legal experts from their organizations and helping them to draft a petroleum law that would enable operations to go ahead.

As a result, the Libyan Petroleum Law of 1955 looked like an equitable piece of legislation that would benefit both parties to any concession agreement, and at a quick glance its provisions did not seem much different from those of the old Middle East concessions.* Profits from the concessions would be subject to a tax of 50 per cent, and a royalty of 12½ per cent on every barrel of oil would be treated as a partial payment of the tax on profits. But the Libyans were to discover that they had also accepted some fine print the significance of which they did not appreciate until it began to affect them. Written into the law by the American experts were two provisions which were especially favourable to their operations. The first was a depreciation allowance, which allowed the concessionaires to deduct a charge of 20 per cent on all physical assets acquired before production. (They also had a choice of a 20 per cent amortization of all their expenditures before production began, or a depletion allowance of 25 per cent of their gross income.) The second provision tied Libya's oil royalties not to the posted price – as it did in the rest of the Middle East – but to the price which the oil secured on the market. Since the posted price, at Arab and Iranian insistence, stayed rigid even when the market price dipped,† this provision was to cost Libya many millions of dollars in the years to come.

The moment the new law was promulgated, the rush began. It made operations especially attractive to independent oil companies, for the lack of a posted price provision meant that they could sell their product at cut rates and thus get a hold in

---

* Though not the new ones.

† As has been recounted in these pages, the oil companies only lowered the posted price once, and Arab anger was such that they never dared do it again. Since that time the posted price has always been much higher than the market price, and it is on the posted price that OPEC members' income is calculated.

the markets still largely controlled by the major companies. Within a few months, fourteen different applicants, including six of the major groups,* had been granted forty-seven different concessions. Esso International was first to make a big strike of oil in their exclusive concession, Area Number 20, and they were followed shortly afterwards by an independent group called Oasis, which was a combination of three US companies: Marathon, Continental and Amerada. Soon it was evident that Libya was about to become one of the most prolific oil countries in the world, and King Idriss and his ministers realized belatedly that the oil companies had pulled a fast one on them, and that they were being short-changed by their own petroleum law.

In 1961, in an attempt to rectify matters, the Libyans enacted an amendment to the petroleum act, which pegged the oil companies' taxable income to the posted prices, but they also allowed a clause to be inserted adding that 'marketing expenses' could be knocked off the new tax rate. Once more, they found themselves victims of some fast legal footwork. The Libyans had meant to legislate the amount allowed for 'marketing expenses' at two per cent, but somehow that figure never got into the amendment. Soon the companies – or a good many of them – were including whacking great amounts for all the rebates which they were making in the markets in order to undercut the majors and sell their oil. And when this loophole was plugged, another amendment got itself voted stipulating that no amendment or repeal of the regulations would affect the existing contractual rights of the concession holders without their consent.

Why did Libya allow itself to go on being outsmarted by the exploiters of its oil wealth?

Ignorance and innocence were not the only explanations. King Idriss was a simple-minded Senussi who had grown old waiting for the liberation of his people, and now much of the fighting strength had gone out of him. His principal pleasure was bringing his young tribal wife to a small palace he had built at Tobruk, where they relaxed and bathed. His only interest in oil was a negative one; he became enraged when

---

* Esso International (a subsidiary of Standard of New Jersey), Shell, Standard of California, BP, Mobil and Texaco.

crude being loaded into tankers along the coast from his palace spilled and drifted in to pollute the beaches.

He had, however, allowed himself to be surrounded by ministers whose venality knew few bounds, and the habit of corruption had permeated most ranks of Libyan officialdom. There is little doubt that much of the loose legislation which cost Libya so much in oil revenues during its early years was due to the large sums of money which changed hands between certain of the independent oil companies and the oil ministry. The major companies did their share of bribery and corruption – they maintained that they had to in order to get any business done at all – but behaved on the whole with much greater circumspection than the independents. Some would say that, in view of their global agreements, they had to; but the fact remains that Esso International, for instance, the moment the Libyan government pointed out the inadequacies of the petroleum law, began to sell its Libyan oil at Middle East posted prices (about $2.16 a barrel) and paid taxes accordingly, amounting to about 90 cents a barrel. On the other hand, many an independent group went on selling its oil at cut-rate prices (about $1.55 a barrel) and paid less than 30 cents a barrel tax.

The major companies could not say so officially (since they did not recognize the organization themselves) but they realized that membership of OPEC was the only hope that Libya had of solving its problems and ending the anomalies in oil payments. With OPEC behind them, the Libyans could talk from strength. But each time OPEC approached the government, one of the independents opened a new private bank account in Switzerland for a Libyan oil official, and that official used his influence to veto OPEC membership. The independents had a case for opposing any legislation that would bring Libya's taxes into line with those of the Gulf states and Iran, and during a conference with the Libyan premier, Husain Maziq, they argued that OPEC membership would not only harm the interests of the independents but would cut into Libya's global oil sales as well.

They pointed out a conflict of interests between the independents and the majors and between the major companies and Libya itself [writes George W. Stocking]. They observed that the majors' interest in producing relatively high-cost Libyan crude was slight compared with low-cost Arabian crude. The increase in their Libyan

costs, which the new royalty and tax provisions would exert on
their total average costs, would be slight, but it would have a disastrous effect on the independents. They suggested that this would not
be displeasing to the majors, who would welcome an opportunity
to buy them out at bargain prices. And they argued that once the
independents had disappeared from the Libyan scene, the majors
would have a cost incentive to increase Persian Gulf output at
Libyan expense.[3]

In other words, they maintained that any legislation which
altered the present pricing arrangements would result in 'a
short-run gain and a long-term loss'.

But then the independents made a mistake in psychology.
Convinced of the validity of their arguments, confident that
they had bought the support of the ministers concerned, they
then let it be known that they would not accept any new
legislation anyway. It was an old story in the oil business, and
Howard Page of Standard of Jersey must have smiled cynically
when he heard about it. Because what the independent
companies said was that the Libyan government could pass any
new laws it liked, but so far as they were concerned a contract
was a contract. They would refuse to accept any laws which
tried to break it.

This was too much even for the supine and corrupt members
of King Idriss's administration, and news of the independents'
arrogant attitude was leaked to the Egyptian and Arabian
press, which began to talk about 'colonialist exploiters' who
'use the USAAF base at Wheelus to bolster their outrageous
demands.'[4] Shortly afterwards, Libya announced that it was
joining OPEC at last. Prime Minister Maziq issued a statement
in which he said: 'If they [the independents] continue this
attitude, we shall be compelled to issue new legislation which
will oblige them to abide by the new system [of taxation].'[5]

That legislation, he then made clear, would be particularly
drastic, and it would be enforced if necessary by Libyan
arms. Any independent company which refused to pay the new
taxes would have its oil exports prohibited and its installations
taken over until payment had been made. To bolster Libya's
threat, OPEC hurried to the aid of its new member. In a resolution passed at an emergency meeting in Vienna, the member
countries pledged themselves to refuse any oil rights whatso-

ever 'to any company or the subsidiary of any company refusing to comply with Libya's new oil policies'.[6]

With some prodding from Washington (where the government made it clear that the independents were on their own in the event of trouble) the companies came into line and accepted the new tax rates. Libya was now a member in good standing of the Organization of Petroleum Exporting Countries, and would henceforth reap the benefits of that body's joint negotiations.

In the meantime, OPEC experts estimated that the Libyans had lost something like $100,000,000 in taxes as a result of the petroleum act. That figure would be remembered by the young militants who were beginning to be active in student and army circles. And one day they would get it back – with interest.

Not that joining OPEC provided an instant cure for Libya's chronic condition of maladministration and fiscal dishonesty. It was still the land of 'the date palm and the greasy palm', as one American described it. In 1965, the Libyan government announced the opening of new areas in Tripolitania for oil development, and from all parts of the world the oil companies came rushing in with their bids. Despite their complaints that Libya's new rules would drive them out of business, most of the independents already operating inside the country were among the bidders.

To the astonishment of the outside world, but not to those who listened to gossip in Tripoli, two of the choicest territories went not to any of the established companies but to a total outsider, Occidental Petroleum Company of California. The company had been taken over some seven years earlier by a formidable entrepreneur named Doctor Armand Hammer, whose expertise as a financial manipulator can be judged by the fact the while still in his twenties he had gone to Russia from the US and managed to make a fortune as a trade go-between in the Soviet Union. In America he had become involved in a wide spectrum of activities exemplifying the breadth of his tastes and interests, for they ranged from hard liquor to op art (he owned a whisky company and the Hammer Galleries in New York), and by 1957 he had bought up a modest California company named Occidental and began his preparations to break into the international petroleum market.

Doctor Hammer blandly ascribed his good fortune in winning the new Libyan concessions to the meticulous care with which Occidental's bid had been prepared, and to the fact that he had paid the minutest attention to Libyan susceptibilities and their nascent pride by tying up the bid in ribbons fashioned in Libya's national colours. But subsequent developments led the less naive to believe that the success of the Occidental bids was due less to the colour of the ribbons round the documents than to the size of the dollar bills which went with them.

As the *Wall Street Journal* was to put it later:

Some little-noticed documents on file in federal district court here [New York] in connection with a suit against Occidental by Allen and Co. show some of the influences that may have been brought to bear in Occidental's winning of its bonanza. Involved are an agreement by Occidental to pay $200,000 to Ferdinand Galic, a bon vivant European business man and promoter; the financing of a documentary film written by Fuad Kabazi, Libya's former oil minister, and said to cost $100,000; alleged payments by Occidental to Taher Ogbi, the company's Libyan representative who became minister of labour and social affairs, and to 'General de Rovin,' a notorious international swindler whose real name is François Fortune Louis Pegulu; and the $100 million breach of contract suit filed against Occidental by Allen and Co., the Wall Street investment banking firm.[7]

Documents deposited with the federal district court in New York in 1972 show that Occidental's fantastic success may well have begun in the summer of 1964 when two colourful characters met in Paris for the first time. One of them called himself General de Rovin and the other was Ferdinand Galic. According to court documents, de Rovin, then sixty-three years old,

was a swindler in Berlin, Vienna and elsewhere in the period prior to World War II; that he had dealings with the Nazis during the War for which a French court later sentenced him to death in absentia; that in the postwar period he travelled about South America and Canada, making a living by passing bad cheques; that he eventually returned to France from Argentina under the phony name of 'de Rovin'; that he was employed by a French firm and promptly squandered its assets; that in February 1970 he was sentenced in absentia to a year in jail by a French court after a conviction of violation of foreign exchange controls.[8]

Galic, a Czech married to an American, and a well-known figure in Paris society, had no knowledge of General de Rovin's gaudy past when they had their first encounter. De Rovin had a proposition. If Galic could find an oil company willing to put up millions of dollars on the project, he had the means, through a highly placed Libyan, of obtaining lucrative concessions in that country. Galic at once telephoned a friend of his, Charles Allen, multimillionaire founder and general partner in Allen and Co., a New York brokerage firm.

'Charley told me, "Give me time, I will look around," ' Galic subsequently recounted in his deposition. In September 1964, Doctor Armand Hammer, Herbert Allen Senior (Charles Allen's brother), Galic and de Rovin had a meeting in Claridges Hotel in London with a Libyan businessman named Taher Ogbi. It soon became clear that Ogbi was the highly-placed Libyan de Rovin had been talking about.

It was at this stage that Fuad Kabazi appeared on the scene. Kabazi was oil minister in the Libyan administration and one of the strong men of the king's government. Introduced to Galic by Taher Ogbi, he said later that he took to the promoter at once. They became 'friends from the first day we met', he said in a deposition deposited with the Federal District court, and he came to admire him 'for his personality and his connections in the world of business'. From that moment on, he kept Galic fully informed of how the government felt about the concessions, and even went to King Idriss himself to press Occidental's case for getting the choicest territories in the promising new field.

The oil minister maintained later that never for a moment did he take the unprecedented steps he did on Occidental's behalf because he was being bribed or corrupted. He knew from the start why Galic was cultivating his acquaintance, he said, and that his main purpose was to secure concessions for Doctor Hammer's organization.

'The whole purpose of his contact [with me] and close relation [with me] was to get this done,' he testified, and though he knew that it was a crime punishable by a jail sentence for a Libyan official to disclose cabinet secrets, he fed Galic with everything that was going on because the Czech represented Occidental and was 'the man authorized to talk and the man to whom I should tell everything'. But he acted because he believed that the actions he was taking were 'for Libya's good', and also, there now seems little doubt, because Galic had completely won him over. He had discovered that Kabazi, a poet and an intellectual, was keenly interested in books and movies, and they talked for hours together about film techniques. Soon they were very close.

'I think every time [Galic] came to Libya, the first person he came to see was myself,' Kabazi testified. 'We also met in Europe. We corresponded approximately every fortnight – the time it takes for two letters to go off and back.'

They used a secret code in which Doctor Hammer of Occidental was referred to as 'marteau' (the French word for hammer), and occasionally their meetings took on a cloak and dagger quality. Once, when the concession bidding was reaching a pregnant stage, the two friends discovered that they were on the same plane together but 'we pretended not to know each other', Kabazi said.

By that time Galic had agreed to put up $100,000 to finance a film which had been written by the oil minister and was to be called *On the Crest of the Dune*. It was to be directed by an Italian, Guido Arata, and Kabazi's brother-in-law was to take ninety per cent of the profits on its earnings.* In his conferences with Kabazi over the script of the film, Arata testified that the minister could not concentrate because he was deeply worried about the problem of securing the concessions, for Occidental. There were influences in the government who wanted other companies to have the choicest blocks, Arata testified. He also had doubts whether Occidental had sufficient money to undertake the vast drilling projects that the concessions would entail, but Galic assured him that Allen and Co. would find the money. And by this time, Arata testified, Kabazi seemed to believe that he had 'commitment' to help Galic to get the concessions for Occidental.

There are many, many interests involved [he told Arata]. Imagine, it's as though there were a large dish filled with all little bones and around this dish are many, many dogs that are trying to edge each other out to grab a hold of the contents of the dish, but in view of the fact that Galic is a dog larger than the rest, he will eat the bone he has asked to eat.[9]

* The film was subsequently made but never shown commercially.

Early in 1965, Allen and Co. discovered that General de Rovin was not all he claimed to be, and felt it their duty to inform Doctor Hammer and Occidental. The deadline for submitting bids for the new Libyan concessions was 29 July 1965, and it was not until two weeks before this date that Occidental sent telegrams to de Rovin and Galic telling them that an agreement Doctor Hammer had signed with them in London in September 1964 was cancelled forthwith. At the same time a telegram reached Allen and Co. telling them that their agreement of December 1964 with Doctor Hammer was also cancelled.

It was too late by that time for Galic and Allen and Co. to work out an arrangement with another oil company. Occidental's bid, gift-wrapped in the Libyan national colours, already recommended by Kabazi and backed by the cabinet ('at the end, when I had the support of the prime minister,' testified the oil minister), was deposited in Tripoli and, to the amazement of the outside world, accepted. Occidental was granted the two choicest blocks in the new field, Concessions No. 102 and 103.

Fuad Kabazi is now in jail in Tripoli, charged with treason by the new Libyan administration. Ferdinand Galic has associated himself with Allen and Co.'s suit for damages against the oil company. Occidental itself thrives. Concessions Nos. 102 and 103 have proved to be two of the most lucrative oilfields in Libya, and they pumped 240,600,000 barrels of high-grade crude in 1970. From a small company which had operating revenue of $800,000 in 1957, when Doctor Hammer took it over, it is now a giant of the industry, with $403,500,000 invested in oil property and average gross earnings of nearly $1,000 million a year.

As the *Wall Street Journal* remarks:

There are, of course, complex dealings in many huge international business arrangements. And no one has even intimated that Occidental did anything illegal in its successful efforts to gain the big oil concession.[10]

Moreover, Occidental contends that it was the company's superior bid which won the concessions, especially since it contained a promise to build an ammonia plant in Libya and to use five per cent of its pretax profits for an agricultural project. To quote the *Wall Street Journal* again:

Nevertheless, Occidental's activities provide an insight into how huge companies sometimes operate in far-off lands.[11]

No doubt more details of those activities will be forthcoming when Allen and Co.'s lawsuit finally comes before the court. But when it does, it seems unlikely that the fast-talking, cosmopolitan General de Rovin will be giving his testimony. According to the records of the federal district court in New York, 'his present whereabouts are unknown'.

PART SEVEN

End of an Epoch?

24 · Boycott

On 4 June 1967, as the Six-Day War began, Gamel Abdul Nasser announced to the world that Britain and the United States had joined with Israel in its attack upon Egypt and Jordan. It was not true, of course, but it was believed by the Arabs, most of whom were already convinced that the British and Americans were tied to Israel by an umbilical cord. Their rage knew no bounds. Their first instinct was to hit back at the Anglo-Americans in the place where it would hurt them most, their oil enterprises in the Middle East. In Kuwait, in Saudi Arabia and along the Persian Gulf, bands of saboteurs assembled with explosives, ready to destroy the big companies' installations once and for all.

Fortunately for everybody, as it turned out, Iraq announced on the night of 4 June that it was taking control of IPC's operations and was shutting down all wells, in order to deny petroleum to the 'imperialist aggressors'. Foreign ministers from the other Arab countries (but not Iran) hurried to Baghdad for conferences, and announced that they would do likewise. Britons and Americans all over the Middle East could be heard sighing with relief as they watched troops moving in to occupy the oilfields and the refineries. Their orders were to stop the flow of oil, but their mere presence would prevent the sabotage of the installations.

By this speedy action, the Arab oil states prevented permanent damage being done to the oilfields, but the boycott into which they had been coerced by Nasser's inaccurate propaganda hit them a body blow in their economy. As they all discovered within a few days of the oil flow ceasing, none of them had the financial reserves to be able to carry on without the aid of their revenues from the companies. Saudi Arabia was the first to feel the pain of deprivation most acutely, and on 12 June, less than a week after the declaration of the boycott, King Faisal was informed by his finance minister that there was no more money in the till and Aramco, for once, was unable to help.

'We must start up the oilwells again,' said the king. 'We know now that Nasser was lying.'

'No Arab will believe that,' his minister said. 'We will be accused of sabotaging him in his hour of defeat.'

'How can we do any worse to him than he has done to himself?' asked the king.[1]

It was at this point that Ahmad Zaki Yamani, the oil minister, made a suggestion. He proposed that Aramco should be told to restart operations, but to agree in writing to withhold all oil from 'those states named by His Majesty's government as having taken part in the aggression on Arab states'.[2] Yamani pointed out that Aramco's oil went principally to the Far East and only the smallest proportion of it went to Britain and the United States, the 'aggressor states', and that while paying lipservice to Arab antipathies this formula would do the least damage to Saudi Arabia's economy. The lifting of the general boycott on these terms was announced on 13 June, but it was not until the end of the month that normal shipments were resumed by Aramco, and already Yamani calculated that Saudi Arabia had lost $30,264,900 in revenue. Kuwait (which estimated its losses at just under $1,000,000 a day) was the next to follow suit, and most of the Arab oil states did likewise quickly afterwards, with the exception of Iraq, which waited until the end of July.

By that time, the Arab oil states had begun to discover the fallibilities of OPEC, which only a few months ago had been thought so strong and united. True, OPEC as a body had not been asked to institute an oil stoppage, but its two principal non-Arab members, Iran and Venezuela, might have been expected to aid their fellow members' boycott by limiting their own operations to the rate of shipments prevailing before the Six-Day War. Instead, they stepped up production to cater for all the shortages their Arab confrères, as well as the closing of the Suez Canal, had caused. The Shah of Iran had been agitating for some time for a considerable increase in the output of oil from the consortium's fields; but the consortium, all of whose member companies had interests in other parts of the Middle

East, had so far resisted a step-up,* on the grounds that taking more from Iran would have meant taking less from Saudi Arabia, Abu Dhabi, Iraq and Kuwait, and thus arousing Arab hostility. Now they did not need to exercise such restraint, since the Arabs had halted oil shipments, anyway. The result was a tremendous rise in Iranian shipments to Britain and Germany, the two main targets of the Arab boycott. Venezuelan production was increased for shipment to Europe (its usual market was principally the United States). At the same time, King Idriss of Libya, who was far from being pro-Egyptian and did not mourn at all over Nasser's defeat, agreed to ban shipments to Britain but not to West Germany, and the increased loads thereupon sent to the Federal Republic were more than enough to divert surplus shipments on to the United Kingdom.

The Arab oil nations met, on 29 August 1967, in Khartoum to discuss the situation, and by now most of them glumly realized that the worst sufferers from the boycott were the oil states themselves. Speakers valiantly hymned the theme of 'oil as a weapon against the imperialists', but privately admitted that economically it had been a disaster for them. From the Arab point of view, the embargo failed because:

1 The US, the principal offender in Arab eyes, was not hurt because of the insignificance of its Middle East imports (about 300,000 barrels a day). On the contrary, the emergency enabled US companies with sizable US, Venezuelan or North African production, to make handsome profits;

2 The international oil industry did an outstanding job of making up the shortfall in supplies from other sources, despite the closure of the Suez Canal. (In this case, the focus of the world oil supply crisis shifted somewhat from the availability of crude oil at the source to a tight situation as regards [tanker] transportation.)

3 No quota ceilings were imposed on liftings of oil approved by the Arab governments concerned, which encouraged overlifting from Mediterranean and other ports. The excess of oil no doubt found its way via West Germany to Britain.

4 There was no uniform interpretation of the coverage of the embargo. The North African countries did not in fact embargo West Germany.

5 Both Britain and West Germany had oil stocks of between 8 and 12 weeks which, with rationing, could be made to last them 5 to 6

* Though they did sell some extra oil to NIOC (the National Iranian Oil Company).

months, even assuming that emergency supplies were unavailable. That is not to say that Britain and the rest of Europe got off lightly, but only that such discomfiture as they suffered was attributable more to the closure of the Suez Canal than to the partial embargo.[3]

The whole affair had been badly handled, everyone admitted. 'Injudiciously used, the oil weapon loses much if not all of its importance and effectiveness,' declared Ahmad Zaki Yamani. 'If we do not use it properly, we are behaving like someone who fires a bullet into the air, missing the enemy and allowing it to rebound on himself.'[4] As a sop to the pro-Nasser mobs, Saudi Arabia, Libya and Kuwait agreed to pay a yearly subsidy to Egypt and Jordan of £135,000,000 ($378,000,000), but all of them fervently hoped that a Middle East demagogue would never get them into such an unhappy and costly situation again.

Unfortunately, it soon seemed only too likely.

One night in October 1967, some four months after the end of the Arab-Israeli War, a taxicab drew up outside a restaurant called Les Ambassadeurs, in Park Lane, London, and four young men got out and went inside. Les Ambassadeurs is a private club much used by film stars, newspaper columnists, and publicity men and it has a gaming-room upstairs where card games are played for stakes high enough to seem impressive, even by Monte Carlo or Las Vegas standards. The owner at the time was a burly ex-Polish army sergeant named John Mills who liked to keep the club what he termed 'exclusive', and young men who were not recognizably famous and did not appear to be millionaires – and this quartet did not – were apt to be turned away.

Two of the young men, however, were Arabs and one of them, moreover, as his passport showed, was a sheikh from a trucial emirate in the Persian Gulf. Many a London gambling promoter's fortune has been made by being polite to sheikhs, and the four young men were promptly allowed inside. It was much later in the evening that they found themselves upstairs in the gaming-room, where they wandered around, watching the gamblers at play. Suddenly, the second Arab, a good-looking but awkward-moving Libyan, halted at one of the tables and said:

'But I know this man! Who is it he plays with?'

One of the Englishmen with him explained that the player he was asking about was a well-known Greek shipowner. The swarthy man with whom he was playing had been recognized by the Libyan as a fellow countryman, one of the king's closest advisers. For the next hour he watched them, and it was an hour that was to change his life. In the time he stood there, the two men between them appeared to have lost something over £150,000, and all the time they were doing so, the young Libyan was staring down at them, his body rigid with tension, his eyes burning. When his companions tried to draw him away, he shook off their hands.

'Let me watch them!' he said, fiercely. And then, in Arabic: 'Carrion, carrion! First they rob us – then this is what they do with the money!'

It was the first – and almost certainly the last – time that the Libyan had ever been in a London gambling club, and it confirmed every opinion he had formed since arriving on a military course in England of the decadence of the country where he was learning to be a soldier. But it was the sight of the Greek and his fellow Libyan, gambling away thousands of pounds on the turn of a card, which excited his hatred, for these were the two elements – Greek shipowners and Libyan ministers – who he believed had profiteered most at the expense of the Arabs during the Six-Day War. And one of these days, the young Libyan vowed, he was going to do something about them.[5] His name was Mouammar al Gaddafi, and two years later he would be master of his country.

It is true that the Greek shipowners did well out of the Six-Day War and the closure of the Suez Canal which resulted, but no better than any other owner with a tanker for hire. The war came just too soon for the big oil companies to take delivery of the giant tankers which they had ordered shortly after the canal had been closed down for the first time, during the Suez crisis of 1956, and they were still forced to rely on the charterers. One of the Greeks, Basil Mavroleon, admitted that he was hiring his tankers to the oil companies in the Gulf – for the long journey round the Cape of Good Hope which the canal closure made necessary – at £5,000 a day on an 80,000-ton vessel. A Norwegian, Sigval Bergeson, was chartering one of his 80,000-ton vessels almost simultaneously to Shell for £1,000,000 for two journeys to and from the Gulf.[6] But

Gaddafi's spleen was misplaced. As Henri Deterding once said, the tanker business is 'the biggest floating crap game in the world', and a Greek was just as much a gambler when he hired out his ships as when he played for high stakes in a London night club. He was just as likely to lose his shirt on either game, and he could hardly be blamed for chartering his tankers at £7 ($17.50) a ton during a crisis – the rate prevailing immediately after the Six-Day War – if the going price was likely to drop to 45 pence ($1.20) during a slump, as it did in 1972. By that time so many new tankers would have been launched that many a Greek or Norwegian charterer would be forced to lay up his ships or try to adapt them for grain shipping.

But Gaddafi's rage over his fellow countrymen's excesses was only too well justified. Libya by 1967 had become an Arabian Klondyke, with all the attendant evils that such a description implies. Tripoli was filled with wheelers and dealers from all parts of the Western world, accompanied by the cheapjacks who tug at their purse strings. Ministers built themselves luxurious villas, opened secret bank accounts in Switzerland, and went to London to whore and gamble. But there were no schools, few hospitals, and urchins hung around the streets, shoeless and in rags.

Whether King Idriss was aware of the corruption going on around him is questionable; certainly he was too old, too weak and too complaisant to do anything about it. To young Libyan army militants like Gaddafi, who were pro-Nasscrites to a man, his most grievous sin was his reluctance to give all-out support to the Egyptian leader during the Six-Day War. Like his fellow monarch, Faisal of Saudi Arabia, King Idriss feared the Egyptian leader for his socialistic and anti-royalist pretensions, and did not mourn overmuch at his defeat at the hands of the Israelis. It is true that during hostilities and for a time afterwards Libya halted the shipment of oil, but that was because King Idriss did not have much choice; the dockers went on strike and oilfield workers prevented any more crude flowing through the pipelines. To get them back to work and to demonstrate his 'solidarity' with his neighbour, the king agreed to join Kuwait and Saudi Arabia in compensating Egypt for its war losses, but his ministers pointed out that the millions of dollars involved in this subvention could only be provided if the oilfields started up again.

The workers returned. The oil companies were told to raise their rate of production to take care of the increased demand which Libya expected in view of the dislocation of oil supplies from the Persian Gulf caused by the Suez Canal closure. At the same time, they were ordered to raise the posted price of Libyan oil by 80 US cents a barrel.

There are three factors in the present situation which call for a review and adjustment of posted prices [the government announced]. 1 The world-wide increase in prices of petroleum and petroleum products, particularly in European markets, caused by the Zionist aggression. 2 The more favourable geographical position of Libya for the supply of these markets in comparison with the Persian Gulf owing to the closure of the Suez Canal. 3 The rise in freight rates which increases the favourable situation and demand for Libyan oil compared with oil from the Persian Gulf.[7]

From that moment, the boom was on. In the year that followed Libya's revenue from oil-production (and she got precious little from anything else) almost doubled as the demands from a petroleum-starved Europe began pouring in. Not much of the increased revenue found its way into funds for the health and welfare of the Libyan people, and large parts of it were siphoned off into the pockets of the king's ministers, and their hangers-on. One Libyan family alone, now safely in exile, was said to have made $4,000,000 in the 1967–9 oil boom. Only too well aware of the venality of their masters, Libyan civil servants, police, customs men, petty officials, got in on the act by demanding higher and higher bribes for their services. 'Gimme Gimme Land' was the new name for Libya among the oilmen.

To the young army captain, Mouammar al Gaddafi, the spectacle was a sickening one. Until the age of sixteen, he had wandered the Tripolitanian desert with his family of Senussi tribesmen, and a goatskin tent had been his home. Cities were decadent places, in any case, in his eyes, but what he particularly despised were the growing evidences on every side in Libya now of what he considered to be the degrading smear of effete Western civilization. Gaddafi was not only pro-Nasser and socialistically-minded, but he was also a puritanical Muslim to a degree approaching fanaticism. Among his roster of heroes only two non-Arabs figured, Abraham Lincoln and

General Bernard Montgomery* – the latter for his ascetic private life and soldierly qualities – but otherwise he loathed the standards of modern Western civilization, and he once confessed that Arabs who aped European ways, associated with Western women, gambled or drank, 'made him vomit'.

In 1969, when the Libyan oil bonanza was riding high, there occurred an event which must have considerably turned the young Libyan's stomach. In the spring of that year Occidental Petroleum Company completed construction of a pipeline from its oilfield in Tripolitania to the port of Sirte, on the Mediterranean. By this time Occidental was in the money – the Six-Day War had been a godsend to the company – and the relationship between its president, Doctor Armand Hammer, and the king, as well as his ministers, was one of cordiality and mutual admiration.

King Idriss consented to perform the opening ceremony of the new pipeline and a guard of honour was picked from the Libyan army for the occasion. Captain Gaddafi was one of the officers chosen. It was quite an occasion. Dr Hammer brought members of his board and friends from the US. All the government ministers came, including Oil Minister Kabazi – though his much-admired friend, Galic, was not present – and the throng of guests was ornamented by the presence of quite a few female members of Tripoli's Italian colony. The young Libyan captain† had a close-up view of his countrymen, arriving in their Rolls-Royces and Cadillacs, drinking champagne, smoking fat cigars, and fawning not only over their American guests but also over the Italians who had once turned Libyans into semi-slaves. The sight appears to have considerably intensified his feelings, and convinced him that his country was rapidly becoming a sink of Babylonian iniquity. He went back to his barracks at Bab Azizia, outside Tripoli, and talked long into the night with a clique of young associates. A few weeks later, he had convinced them that a revolution was feasible, and had even written the slogan with which he would rally the people: 'Poor and barefooted, but with Nasser!'

On 1 September 1969, while King Idriss was on an official

---

* Commander of the British Eighth Army in Libya during World War II, and subsequently British c-in-c in Europe.

† He became a colonel later.

visit to Turkey, Gaddafi and his group quietly moved companies of faithful troops to strategic positions throughout the kingdom, and then took over the regime in a bloodless coup d'etat. Within hours, several ministers (including Kabazi and the gambler whose excesses had so shocked Gaddafi in London) were under arrest, and the desert kingdom which many Arabs called 'Camel with the Feet of Gold' was no more.

A few nights later, a young man wearing a dark cape over his shoulders slipped into a well-known Tripoli cabaret and stood at the bar, sipping an orange juice and staring at the crowded tables of guests loudly clapping a belly-dancer performing on the dance floor. Suddenly he whipped off his cape, revealing his army uniform, and began to blow on a whistle. Troops burst into the room, and a hundred and fifty guests and cabaret girls were shepherded into trucks and driven off to jail.

Next day all cabarets were closed down in Libya, never to open again. Newspapers covered their front pages with denunciations of their immorality, and of the decadence that contact with the West was bringing to the Libyans.

Not long afterwards, the same young man, dressed this time in a simple Senussi gallabiah, stood in the attendance room of a Tripoli hospital and waited while a group of nurses and doctors chatted together. Finally, timidly, he approached them and asked for attention.

'Come back tomorrow,' said one of the nurses, abruptly.

'When I come back tomorrow,' the young man said, 'you will no longer be here. You will be in jail.'

Soon all Libya was talking about the nocturnal peregrinations of the young wolf of the Libyan revolution. Gaddafi had become a modern Haroun al Raschid, liable to appear anywhere among his people, watching, listening, and then acting. Slowly, at first, but inexorably, the purge of Western influences began. Each time British and American oilmen came in on leave to Tripoli, they found more evidence that things were changing. Bars had closed down and liquor shops. Street names were in Arabic only. Urchins no longer importuned for money, and hotel servants were no longer subservient – or willing to give service. You could almost smell the xenophobia in the air. And the new rectitude of the government was tangible. An Esso International official came back to his office to declare in astonishment: 'I actually got a paper signed without having to pay baksheesh for it.'

But Gaddafi would soon be finding other ways of making the oil companies pay. First, however, he busied himself with preparations for his first big challenge to the Western influences which had dominated Libya for so long. A few weeks after taking over, once the people had been prepared and the propaganda groundwork laid, he was ready. He ordered the US Air Force to quit its base at Wheelus airfield and get out of the country. The Arab world waited, and the militants held their breath. How would the United States react to this impudent challenge from a group of upstart young men to their strategic policy in the Middle East? From Saudi Arabia, Faisal is said to have told the exiled King Idriss: 'Don't worry. You will soon be back. The Americans will see to it. Those misguided young men have gone too far.'[8]

But neither he nor many another Arab incumbent realized how things had changed for the big powers in their confrontations with the new regimes which were rising out of the ashes of the Six-Day War. From now on, arrogance and defiance were going to get the Arabs everywhere, and the era of subservience was over. The US Air Force packed up its supplies and obediently flew away, to be followed not long afterwards by the British from their bases in Tobruk and Benghazi. True, the British did not go quite so willingly, but a few riots in the streets, some Britons beaten up and nurses raped, speedily brought the situation to the point where the British had either to evacuate – or bring in the modern equivalent of a gunboat. But, as Suez had proved in 1956, the days of gunboat diplomacy were over. If the United States was in no position to defy the exigencies of the revolutionaries of Libya, how could a weakened Britain dare to do so?

The foreign troops departed and down came the Pepsi Cola posters and all the other signs of Coca-colonization.* How long, the oilmen began to wonder, would it be before the young wolf got around to them?

In the home offices of the big oil companies, in fact, they were

---

* Coca Cola itself was already banned by all Arab countries in retaliation for its commercial activities in Israel. The Ford Motor Company and Alka Seltzer were also under Arab proscription for the same reason.

not worrying too much about the sword of Damocles now hanging over their operations in Libya. If it had been 1967 or 1968, the threat of Gaddafi's interference in the oilfields would have given them ulcers, because it was Libya's oil which was keeping Europe's wheels turning in the aftermath of the Six-Day War. But by 1970, the Anglo-Americans had the situation in hand. The giant tankers ordered in the aftermath of the 1956 Suez crisis were now beginning to come down the slipways, and freight charges from the Persian Gulf were starting to drop. Stocks of fuel were building, and soon the shortage would be over.

In the circumstances, what had they to fear? The Libyan leader was in no position to hold them to ransom. It was a question of facing any demands he made by a united front, and refusing to knuckle under to his exigencies. When it came to the crunch, it would be Libya that would suffer most. The oil companies had their alternative sources of supply – in the Gulf, in Iran, in Venezuela and in their reserve stocks – with which they could keep their customers supplied. Even if the Libyan fields came to a complete halt, none of them need lose a cent in sales or profits. Whereas Gaddafi would lose his oil revenues just at the moment when he needed them to replace the millions which the ministers he had usurped had smuggled out of the country.

Veterans of tough negotiations in Iran, Iraq, Venezuela, with countries ten times the size of Libya and dictators twice his age, the oilmen were convinced they could handle the young colonel and his band of militants. In the eyes of many an oilman they were just a bunch of naive fanatics with sand running out of their ears. The word went round to stand up to the new regime, and not take any nonsense.

As they were to discover to their cost, Mouammar al Gaddafi is not so naive as he looks and sounds, and in his first encounter with the major companies in Libya he outwitted them. Not for the first or the last time, he consulted his second-in-command in the Revolutionary Council on the strategy to be adopted, and Major Abdessalam Jalloud did not let him down. Among this strange pack of Islam fanatics, dedicated to the precepts of the Koran, the glory of the Arabs, and the downfall of the Zionists, Jalloud stuck out like a pine tree in a desert. He was a towns-man rather than a child of the desert, and there were times when he was known to sigh over what his fellow rebels had done to Tripoli in pursuit of faith and asceticism. A sojourn in the West, far from giving him a revulsion against modern civilization, as it had in the case of Gaddafi, had given him a taste for European culture, a liking for jazz, and an avid taste for good food, wine and intelligent female companionship. Some of the beautiful young members of the Italian colonies in Tripoli and Benghazi had come to know him well, and he had continued his companionship with them some weeks after the revolution, and he had given them up only when it became evident that Gaddafi would make them the next target for expulsion. When the militants first took over the reins of government, they all lived together in the barracks at Bab Azizia, an arrangement that Jalloud found constricting and monastic*, and shortly afterwards he moved to a villa some distance out of Tripoli which had been taken over from a departed minister, and was luxurious in the extreme. The young major did not enjoy its sybaritic pleasures for long, because about a week later he drove out one morning to discover that sign-painters had been at work during the night, and on the garden wall of the villa had been painted: 'Who bribed you with this villa?'

He did not need to ask who had ordered the painting of the insulting graffiti, and next day he said goodbye to his wine cellar, his beach and swimpool, his air-conditioned bedroom and his girl-friends, and moved back to the communal life of the military barracks. It is a measure of his popularity with Gaddafi, and of his usefulness to the new regime, that nothing was ever said to him about his attempt to break out of the monastery. Plenty of Libyans had ended up in jail for doing far less.

Jalloud's conviviality was not expended simply in the pursuit of personal pleasure; his general air of bonhomie made him popular with oilmen and he picked up information about the way the oil companies were thinking which the junta might otherwise have missed. He knew they had expected Gaddafi to show his contempt for them by nationalizing the oil operations in Libya, and it was certainly a move that the young militant leader was known to favour. Jalloud had persuaded him not to

---

* Though, in fact, none of them has ever deprived himself of wives or other female appendages.

make that mistake – for, of course, Libya would never at that time have been able to run the oilfields on its own. He urged him to seek financial rewards instead, and advised him to make the companies pay an immediate rise of 50 cents in the posted price of Libyan oil and an increase from 50 to 60 per cent in Libya's share of the sales. Gaddafi wanted to make it a straight $1 raise and turn the confrontation into a trial of strength.

'We must show we are the masters here,' he said.

Jalloud told him he had already demonstrated that well enough by kicking out the American and British armies, and warned him against empty political victories at a time when Libya needed something more important: money. What Jalloud saw as his main problem was how to split the ranks of the oil companies. If they remained united, he knew he could not beat them. But if one of them broke the others would have to break too.

Word spread through the oilfields that the crunch was coming, and the major companies got ready to face up to it, united in their resolve not to give way. They waited for the summons. It did not come. Jalloud had called in the representative of one company only, Occidental Petroleum. Of all the independents operating in Libya, Occidental was the most vulnerable to an ultimatum. It had no alternative sources of supply. Its fortunes had been based on its success in Libya, and $450,000,000 of its profits had been invested in new developments inside the country. It could not afford to have its operations closed down. It could not even face a cutback in production rates, and any hesitation it may have had about accepting the Libyan demands was hastily forgotten when troops moved into the oilfields and ordered an immediate cut in Occidental's output by 30 per cent.

On 14 September 1970, the company announced that it was accepting a compromise agreement with the Libyan government and immediately lifting its posted price of oil by 30 US cents per barrel, and increasing its tax level from 50 to 58 per cent. All the other independents hurriedly announced that they were following suit. By the end of the month the major oil companies had come into line. The united front disintegrated, and one by one the concessionaires were summoned to sign new agreements.

The major companies had been outsmarted, and it was a first victory to the Council of the Revolution.

As one cynical oilman put it : 'At this rate, Gaddafi will soon be getting ideas above his nation.'

25 · Tidying Up

The British are a conscientious people and when they give up an Empire and the occupation of somone else's country they always try to leave it clean and tidy. They are not particularly proud of the mess they left behind them when they quit India in 1947 and Palestine a year later, but they had the excuse on both those occasions that the departure came at the end of a long and exhausting war, and that, in any case, the inhabitants of the lands they were leaving had made it abundantly clear that they preferred any dirty linen which the British might leave behind to any prolongation of their presence. Since those two regretted lapses, Whitehall had contrived to do better, and the lowering of the union jack over Malaya, Singapore, the West Indies and the several states of East and West Africa had been carried out over handshakes and mutual expressions of goodwill. More important, reasonably stable administrations had been left behind to carry on the business of government.

Now it was the turn of the Trucial States of the Persian Gulf to gain their independence, and since Britain would be retaining in the territories some of its most valuable overseas investments – mainly its petroleum concessions – it was considered particularly important to hand over the reins of government to regimes strong enough to control an unruly and backward populace, but supple and forward-looking enough to cope with its demands for improvement.

There were some heavy sighs and doubtful shakings of the head from old colonial civil servants and true-blue Britons when it was announced that Britain would quit the Gulf on 1 December 1971. For more than a hundred years, the Gulf had been a British lake controlling the route to the Far East and India, and for most of that time Britain would rather have fought a war than allowed any other nation to gain military influence there. In 1900, when he was viceroy of India, Lord Curzon and his American wife, Mary, sailed into the Gulf in Her Majesty's ship *Hardinge* and descended upon the Sultan of Muscat and Oman, who, the viceroy had heard, was thinking of accepting a large bribe from the Russians to allow them to establish a coaling station in the area. Curzon had already expressed the view that any Briton who allowed a foreign power to infiltrate into this vital area of British influence should be impeached and hanged. Now, with the guns of Her Majesty's ships trained on them, Curzon and his lady stepped ashore to sip sherbet and exchange welcoming speeches with the sultan.

The sultan referred in his remarks to Mary Curzon as a great and charming beauty, a veritable 'pearl'. Curzon replied by saying that the Persian Gulf was also a pearl, and, with a certain emphasis in his voice, it was a pearl which Britain considered beyond price. The point was taken and more sherbet was drunk.

The morning after we left Muscat, [wrote Mary Curzon in her diary] we cruised all day among the islands and fiords of the most barren description. The land is uninhabited save for a few fishermen, and its main interest is a strategic one, as the bays and inlets afford anchorage for a fleet, and as the land is No Man's Land, Russia or France could take advantage of the harbourage in the event of war.[1]

It must never be allowed to happen, Curzon vowed. And during his lifetime it did not. But times had changed. Britain no longer had the military strength nor the political influence to be able to maintain itself in the Gulf against local hostility, and there was plenty of that among the Arabs since the Suez crisis. Experience in Aden – where a raffish set of left-wing brigands had seized power when Britain left – had taught the British that staying too long could produce a more radical local government than if they departed sooner and more willingly. Aden was a mistake to be avoided in the Gulf, and they dispatched there their best Arabist troubleshooter to make sure that when they quit no wild revolutionaries would immediately take over. His name was Sir William Luce.

British foreign service officers come in all shapes and sizes, but in some ways they are always the same. Luce was a chip off the same block which had produced Sir Percy Cox, whose stern paternal confabulations with the Arab leaders had shaped and delimited the boundaries of Arabia after World War I. Son of an admiral, down from Cambridge with a good degree and a 'blue' for rugby, a devotee of riding, shooting and sailing (he still gives the three as his recreations in *Who's Who*), Luce had climbed the ladder of the colonial service by way of the Sudan, spoke fluent Arabic, and had been dealing with Muslim tribes of one sort or another for most of his adult life. From 1960–66 he had been British political resident in the Gulf, with his headquarters in Bahrain, and during that time he had followed the time-honoured British policy, where Arabs were concerned, of never interfering with the local rulers except to encourage rivalries between them, these rivalries being sufficient to keep them divided and therefore not likely to combine together to kick the British out.

Now suddenly he was called upon to reverse his role. With Britain departing, taking with her the gunboats and troops with which she had protected both the sheikhs and the petroleum routes to Europe and the Far East, the small states along the Arabian shore must be left in a condition to defend themselves. Not against attacks from each other, however, but from incursions from their more powerful neighbours, or from the revolutionary elements which were starting to fan across the Arabian deserts on the heels of the British withdrawal. Therefore, instead of encouraging the Gulf states to glower and snarl at each other across their several frontiers, Luce must now try to persuade them to forget their rivalries and come together. *Federation* was the magic word he brought to the sheikhs, a word that would work as many miracles for them as *abracadabra!* If only they would combine together in a federation of Arabian states, with their own joint army, air force and navy, they would be able to withstand assaults upon their independence from all but the major powers. Most important of all, they would be able to protect their oilfields from infiltration and sabotage.

To emphasize the support that Britain was prepared to give to such a federation, Sir William let the various rulers know that his government was prepared to sign treaties of friendship with them all, and to help them form combined defence forces, Britain would assign to them the locally-raised levies which had been used to maintain colonial order. Their British officers would be left behind with them, though they would, of course, have first resigned their British commissions and would sign on as 'advisers'.

While the emirs and sheikhs were brooding over Sir William's offer, the British envoy was planning another move which was the antithesis of all he had been taught as a good colonial servant: he was preparing to interfere in the domestic affairs of one of the Gulf states. For if a federation did come about, the combined states must have administrations which could rely on the support of the majority of their subjects. By 1970 it could be said that most of the British-controlled Gulf states – especially three of the four key states that mattered, Bahrain, Abu Dhabi and Dubai* – had rulers whose regimes were reasonably stable and against whom a revolt was unlikely. But to the south of Abu Dhabi and Dubai, where her frontiers were inextricably mixed up with those of the members of the proposed federation, was the sultanate of Oman. Oman was not being asked to become a member of the federation, because it was so large and unwieldy that it would have unbalanced the union. But it was so closely involved, geographically, socially and economically, with its neighbours that what happened to it would inevitably affect the others. And Oman was the rotten apple in the barrel. It was a hotbed of misery and discontent and the British believed that the main reason for its unsettled state was the despotism of its ruler, Sultan Said bin Taimur, whose tyrannical methods were fanning the flames of revolt all over the southern deserts of Arabia. Before federation came, and the new independent states were left on their own, something had to be done about Sultan Said bin Taimur. Otherwise the diseases his misrule provoked might spread the infection far beyond the frontiers of Oman.

Oman in 1970 could hardly be described as an oil state in the same class as Abu Dhabi or Kuwait, but the fact remains that its fortunes had improved considerably in recent years. Further north, in Oman proper, a company called Petroleum Development (Oman) Limited† had begun exploiting reasonably productive fields in the desert at Al Huwaisa, Yibal and Natin Fahud and sending oil by pipeline from a collection point at

* Kuwait had been granted its independence by the British in 1961.

† In which Shell held 85 per cent of the shares, Compagnie Française des Petroles 10 per cent and the Gulbenkian interests 5 per cent.

Izki to a tanker-loading port on the Gulf of Oman near the capital city of Muscat. Revenue from this operation was sufficient for PDO to pay Sultan bin Taimur $75,000,000 a year in taxes, and that should have been enough to bring about some amelioration in the lot of the Omani people. But the sultan was a stubborn man and he clung to his belief in a feudal state dedicated to faith in god and the teachings of the Koran. His people not only lacked hospitals and schools, they were also oppressed in spirit by the rigid rules of the theocratic kingdom where what were minor vices or indulgences in other parts of the world were here major crimes punished by death, amputation or flogging.

Young Omanis rich enough to go abroad for their education came back with stories of the wonders of Cairo, Beirut and Europe and their discontent with conditions in their homeland had begun to permeate the youth of the country. They listened avidly to the propaganda broadcasts from South Yemen giving support to the rebels operating in the Jabal Akhdar of Dhofar. They were sympathetic to the rebels because anything was better than the slough of hopelessness and despair which was Oman under the sultan.

On 11 June 1970, members of the sultan's armed forces, the British-officered defence force which kept law and order in the kingdom, were attacked in one of their camps at Izki, near the pipeline collecting point of the PDO. Mines were placed along the barbed wire perimeter of the camp and when they went off a group of guerrillas burst through, throwing hand grenades and aiming bursts of machine-gun fire at the SAF tents. It was dark when the attack was made, and there was great confusion. Fortunately, however, the infiltrators seemed to be just as confused as the defenders, and after a sharp exchange of fire they were all rounded up, and either killed or captured.

The prisoners revealed themselves under subsequent questioning as members of PFLOAG (the Communist guerrilla organization, the Popular Front for the Liberation of Oman and the Arabian Gulf). It was a shock to know that they were operating this far north, for PFLOAG's activities until now had been confined to Dhofar. There were more unpleasant surprises to come. Further questioning at the hands of the sultan's own executioners revealed the whereabouts of several caches of arms smuggled in from Russia and much revolutionary literature, most of it printed in China. Some of the caches were uncovered in Muscat itself. Finally, names of conspirators and secret sympathizers with the rebels came out of the interrogations, and they included officials close to the sultan's court.

The sultan ordered an immediate purge and the foetid, steamy old Portuguese fort in Muscat which he used as a jail was soon full of suspects, most of whom he intended to keep there until they rotted. He was now so alarmed by the extent of the conspiracy that he ordered his own son and heir, Prince Qaboos, to be confined to his house near Muscat. The sultan was taking no chances. He had usurped his own father, and was determined that his son would never get the chance to do the same to him.

But Prince Qaboos had friends. He was an amiable young man whose outlook on life was the complete opposite to that of his father. Though trained at the military college of Sandhurst in England, he had not confined his studies or his observations to military tactics, and he had ideas for Oman which would have outraged the sultan by their liberality and worldliness. He had often discussed these ideas with Wendell Phillips (whose six-shooter he proudly wore), and with the two senior British officers of the SAF, General Graham and Colonel Hugh Oldham. These two soldiers had become admirers of Qaboos and had little doubt that he would make an infinitely better ruler of Oman than his diehard father. At the sultan's orders, the SAF was supposed to keep Prince Qaboos under a tight guard and prevent him from seeing anyone who might conspire with him. But since both Graham and Oldham were officers of the SAF, there was no such restriction upon their own contacts with the prince. From the time of the Izki guerrilla attack, their meetings with Qaboos were almost as frequent as their conferences with the British envoy, Sir William Luce.

Rumours were by now ricocheting around the Gulf that Sultan bin Taimur had decided to abdicate in favour of his son. Who had started the rumours no one could say – though they were strong enough to be repeated in a dispatch from its Cairo correspondent in the Paris newspaper, Le Monde, on 19 July 1970. The British blandly denied having anything to do with them, and Sir William, when approached, pointed out that Britain never interfered in the domestic affairs of the Gulf states. The sultan himself vehemently repudiated any intention of abdicating, and called in General Graham to order him to reinforce the guard on his son's residence. He thereupon departed in a Royal Air Force plane for his palace at Salala, eight hundred miles south of Muscat, where, he imagined, he would be well out of the way of any conspirators. The governor of the province was an old fighting companion, and Said bin Taimur trusted him implicitly.

But on the afternoon of 23 July 1970, his old friend was visiting the island of Misirah, some miles up the coast on the Arabian Sea, where the RAF had a base. Normally he would have turned down the invitation to be a guest of the British airmen, for General Graham arrived in Salala that day for a conference with Colonel Oldham, and on such occasions the governor liked to be present to receive the greetings of the commander-in-chief of the sultan's armed forces. But on this occasion, someone had neglected to inform him that the general was coming. Someone had also neglected to inform the sultan himself of General Graham's presence in the town. The commander of his personal bodyguard, Colonel Turnhill, was, on the other hand, only too well aware of the general's visit, for he was on the airstrip with Colonel Oldham to greet Graham when he touched down in an RAF transport, and the three officers immediately drove away down the coast road out of Salala, on what they later described as a sightseeing tour of the Portuguese forts which dot the coast in these parts. By the time they returned things had changed in Salala.

The climate in and around Salala is one of the most pleasant on the Arabian peninsula, and the sultan's palace catches any cool breeze which comes zephyring in across the Indian Ocean. It stands on the seashore overlooking a series of white beaches and dominates the town, as James Morris describes it, 'like the castle at Windsor or the Mormon Tabernacle in Salt Lake City'. Morris adds:

All these seaside capitals of Eastern Arabia had their big fortress palaces in which the ruler reclined among his dependants and syco-phants, his outlook, like his creature comforts, generally depending upon the size of his oil royalties. Kuwait's was the grandest, Bahrain's the most stylish, Doha's the most horrible, and Salala's probably the nicest. It [is] a long crenellated building surrounded by high walls and complicated by connecting courtyards and alley-

ways. A wide avenue of palm-trees led to the double gate-tower at
its entrance. . . Every Tuesday morning a slim young Englishman
in uniform marched out of the palace after his weekly conference
with the ruler—he was the commander of the Dhofar Force, the
sultan's local army. Muscular well-armed slaves guarded the gate-
way to the palace.[2]

But on this occasion, the commander was away sightseeing
with his general. The guards drowsed in the afternoon sun. At
least they must have been drowsing, for how else could they
have missed and failed to challenge a small body of Omanis in
the uniform of the sultan's defence force who passed under the
gate-tower and jogged at the double (they were wearing running shoes) into the main body of the palace? Said bin Taimur
was in his study, a long room overlooking the sea, a copy of the
Koran in his lap as he squatted on the carpet, his back resting
against an ornately upholstered camel saddle. Two slaves stood
in shadowy corners of the room, watching over him. But none
of them heard the faint shuffle of soft shoes outside nor guessed
that the guard on the door was being overpowered and carried
away.

The great door of the chamber opened and seven men slid
inside. One of them the sultan recognized at once as Braikh,
the young son of his old friend, the governor of Salala, and he
angrily asked him why he had burst in so abruptly. Braikh
replied that he had not come to offer harm to the sultan's
person, but was acting under instructions to take him away. If
he would consent to come willingly and at once, there would be
no trouble. The young man spoke with the utmost respect, and
repeatedly called the sultan by his royal title, Jalalath
(Majesty).[3]

But Said bin Taimur was neither impressed by the intruder's
deference nor frightened by the posse of men standing behind
him, armed to the teeth, daggers in their belts, their fingers on
the triggers of their rifles. As he was to say later, he would have
been ashamed to have given up without a fight, and he had no
intention of doing so. Reaching under the cushions piled alongside the camel saddle, he pulled out a revolver (a six-shooter
of exactly the same type as Wendell Phillips had given to his
son) and opened fire at Braikh. The young man ducked just in
time, and the bullet hit one of the men behind him. At the same
time, the two huge African slaves emerged out of the shadows

of the room, unsheathing the great swords which they wore around their waists. Flailing them above their heads, they waded in to do battle. One crashed to the floor, shot in the stomach, but the other's murderous blade, whirling like a windmill sail, effectively held the intruders back while his master dashed to the end of the room and disappeared behind some curtains.

There followed what might well have been a scene from an old-fashioned film chase. The palace at Salala not only has many corridors leading from building to building, but its walls have concealed in them hidden doors, sliding panels and dark hideyholes in which those inside them can see but not be seen. The sultan utilized them to the full and might not have been discovered for some considerable time had not the young Sheikh Braikh, by this time frantic at the way things were going, started shooting wildly into the panelled walls. Fortunately for him, he scored a hit. From behind one wall came a sudden cry and then a moan for help, and Sultan Said bin Taimur was brought out, bleeding badly from a wound in the shoulder. Patched up by an RAF doctor (who was quickly available), Taimur had sufficiently recovered by evening to consent to abdicate in his son's favour. But to ensure his own safety, he insisted that he would sign the papers only in the presence of Colonel Turnhill, commander of his bodyguard, and with a guarantee of safe conduct from General Graham. The two Britons in question, by this time back from their sightseeing trip, were immediately available to give him the assurances he demanded.

He signed at once, and left for England in an RAF plane the following day. Great was the rejoicing in Oman at his departure. When news reached Muscat of Qaboos's accession, there was dancing in the streets, lights stayed on after dark, the curfew was abandoned, and music from Radio Cairo and Kuwait blared out from radios. Even women came out into the streets and a few of the younger ones dared to lift their veils. All the things which had been forbidden for so long (except, perhaps, the imbibing of alcoholic liquor) were done in the next few days in Muscat and Oman. And to those relatives and progressive Omanis who had been driven into exile by Said bin Taimur's despotism, Qaboos now sent messages telling them to come home, for all was forgiven.

One of the first to send telegrams of felicitation to the new sultan was the chairman of the Shell Trading Company, whose subsidiary, PDO, accompanied it with a message expressing their joy at the 'historic event'. The Foreign Office in London was more restrained and declared that it 'would not dream of embarrassing Sultan Qaboos' by commenting on a purely internal event. Colonel Oldham assured newspaper correspondents now arriving in Muscat that he was 'prepared to swear on my honour that I knew nothing about what was taking place', and General Graham in turn declared that he had been in Salala on the day of the coup 'purely by accident' and had been 'an involuntary witness' of what had taken place. He added: 'But as a good Omani officer, disciplined and respectful of authority, I could not do other than submit to the orders of Sultan Qaboos, to whom I have already given my oath of allegiance.'[4]

The only person who did not make any comment at all was Sir William Luce, the British envoy in the Gulf. But then he did not need to. The rotten apple in the barrel had been discarded. He could now get on with the job of turning the rest into a marketable commodity.

The federation which Sir William and the British had originally envisaged was to consist of nine states in all: Bahrain, Qatar, Abu Dhabi, Dubai, Sharjah, Fujairah, Ajman, Um al Kuwain and Ras al Khaimah. But first Bahrain and then Qatar dropped out, both of them arrogantly confident that they were rich enough and secure enough not to need the federation's protective umbrella. That left seven. Then the ruler of the tiny state of Ras al Khaimah, Sheikh Sakr bin Mohammad al Qasini, blew the wind out of Luce's sails by announcing that he too was staying out.

Sheikh Sakr was (and still is) one of the wiliest characters in Arabia, and almost certainly comes by his combination of guile and truculence by inheritance. His ancestors were responsible for the British installing themselves in the Gulf in the first place, for Ras al Khaimah up until 1820 was the centre of the bloodthirsty buccaneering industry which first gained for this part of Arabia the name of the Pirate Coast. Ras al Khaimah sticks like a sore thumb into the waters of the Gulf and, with two islands which it owns called Greater and Lesser Tunb, it controls the thirty-mile Strait of Hormuz between Arabia and

Iran through which all ships must pass to get from the Gulf to the Indian Ocean. From its ports and islands, the pirate fleets of Sheikh Sakr's ancestors sallied forth to pillage the merchantmen passing through on their way from Iraq, Iran and the Gulf ports of Arabia. When the Royal Navy, exasperated by the pillage and the harassment of its lines of communication with the Indian Empire, at last descended upon Ras al Khaimah they found sixty-three warships and eight hundred smaller vessels anchored there, all of them engaged in buccaneering. The dockside sheds were stuffed with booty, and life in Ras al Khaimah town was gaudy in the extreme.

Sakr, unable to forget the adventurous days of his forebears, dreamed of restoring his country's colourful glories – if not the mayhem that went with them – once his country had succeeded, like his coastal neighbours, in finding oil. He had handed over the concession to the Union Oil Company of California, and never have a bunch of American oilmen worked more dedicatedly to fulfil the sheikh's hopes as well as their own. Surveys of the offshore concessions had revealed 'promising' evidences, and rigs had been brought from California at enormous expense and drilling started. Each day Sheikh Sakr sailed out to see how things were going in one of the units of the Ras al Khaimah navy. The navy was nothing like the once mighty pirate fleet of his ancestors, for it consisted of three rubber dinghies with outboard motors and one small pinnace, but it carried the Ras al Khaimah flag around the Islands of Tunb and flapped it defiantly under the cliffs of mighty Iran, on the far shore. Sheikh Sakr knew quite well that the Shah of Iran, fearful of the blackmail to which the Arabs might one day subject his tankers on their way from Abadan and Kharg Island to the outside oceans, was determined to gain control of the Islands of Tunb for himself. For who controlled the Tunbs controlled the Strait.

Sir William Luce had been to see the Shah of Iran in the course of his negotiations, suggesting that Sheikh Sakr accept the same compromise which he had arranged between the shah and the ruler of Sharjah, which also had a small strategic island in this part of the Gulf. This was Abu Musa, around whose shores were also promising evidences of oil. Sir William had negotiated an agreement whereby the ruler of Sharjah accepted an Iranian military presence on the island, agreed to split both sovereignty over the island and all oil revenues discovered around it with Iran, and would, until such time as oil was found, accept from Iran a yearly subsidy of £1,500,000.

This comfortable compromise could be arranged between Iran and Sheikh Sakr over the Islands of Tunb, Sir William suggested, if only the ruler would give his consent. But no, he would not. The drills were going down into the sea bed off Ras al Khaimah and at any moment the Union Oil Company would hit a rich field of petroleum or gas. Then Ras al Khaimah would be rich and by staying out of the federation, by holding on to the Islands of Tunb, would become a great power in the Gulf, once more to be held in respect by all who sailed through her waters.

There are 54,000 people in Ras al Khaimah, and soon very few of them were unaware of the fact that their ruler had opted for independence, for sovereignty over the islands, and for a gloriously prosperous future, lapped by oil. In the meantime, they lived on the stake-money raked in by the Lebanese croupiers at Ras al Khaimah's casino – the only one in the Gulf – and from the sale of vegetables to neighbouring emirates. And on hope.

In the summer of 1971, those hopes were dashed. John Turk, Union's general manager, drove to the ruler's palace to tell him that 'heavy shoals' of gas had been struck in one location at 17,900 feet, and that 'high quality oil' had been found in another location at 16,000 feet.

'But this is good news,' said Sheikh Sakr.

Unfortunately, however, neither the flow of gas nor of oil had lived up to expectations. Only 2,000 barrels a day were coming out of the well. Union Oil had already spent $15,000,000 on operations off shore, not counting the $200,000 annual subvention which they made to the ruler, and the way ahead looked 'real bumpy', to use Turk's words. Work would continue. Oil would be found and pumped. But the bonanza which everyone had hoped for seemed a long way off.

When news reached Sir William Luce of the waning hopes of Sheikh Sakr, the indefatigable envoy* set off at once for Ras al Khaimah to tell the ruler that it was not too late for him to change his mind. A place in the Union of Arab Emirates (as

---

* He had now been travelling back and forth around the Gulf, and thence to Teheran and London, for over two years.

the new body would be called) was still open to him. So, it was quietly suggested, was a joint arrangement with Iran for the control of the vital Islands of Tunb.

Unfortunately, Arab 'face' was now involved. Sheikh Sakr was a proud man, and he had told his people that their country would be independent. How could he now go to them and say that he had been forced to give way, to knuckle under to rich Abu Dhabi and Dubai in the federation, and to allow Iranian soldiers to establish themselves on the barren shores of the Tunbs? It could not be done. The people would jeer at him for a weakling and a fool. He would have to stand by his decision. There was no other way out.

But there was, of course. When you have able colonial service advisers of the calibre of Sir William Luce at your elbow, there always is an acceptable alternative.

British control of the Trucial States was due to end on 1 December 1971. Until that time, Britain was still technically the 'protecting' power of all the emirates of the Gulf, and bound by treaty to come to the aid of any one of them which might be attacked by a foreign power.

On 30 November, twenty-four hours before the treaty ended, Iranian naval vessels anchored off the islands of Abu Musa, and the Greater and Lesser Tunbs, and troops and marines were ferried ashore. On Abu Musa they were welcomed by a delegation from the ruler of Sharjah, with whom, thanks to Sir William Luce, they had made a prior 'arrangement'. But when the marines went ashore on Greater Tunb – the only one of the two islands with any inhabitants – a Ras al Khaimah police post opened fire, and in the subsequent skirmish three Iranians and one Arab policeman were killed. But a few hours later the islands were flying the Iranian Imperial flag and units of the armed forces were settling in. The Tunbs were in Iranian hands, the Straits of Hormuz were under Iranian control, and the petroleum tankers which passed through them every ten minutes by day and night from the great oil ports of the Gulf would be under Iranian supervision from now on.

Great were the cries of indignation from all the centres of the Arab world. Iran was attacked as an aggressive power seeking to impose her will upon the Arabs. But, to no one's surprise in London, least of all to Sir William Luce, the main barrage of

Arab indignation was turned upon Britain, who was savagely accused of having broken her treaty obligations in the Gulf and failed to come to Ras al Khaimah's aid in her hour of need.

It was all as Sir William and Sheikh Sakr had expected. Far from losing face over the loss of the Tunbs, the ruler of Ras al Khaimah suddenly became the hero of the hour. He had dared to fight in a situation where Britain had cringed and run away. He was hailed by the people not only of Ras al Khaimah but of all the other states of the Gulf. And when, on 2 December 1971, the rulers met in Abu Dhabi to declare themselves formally members of the Union of Arab Emirates, many of the sheikhs expressed deep regret that the 'heroic' Sheikh Sakr was not among them.

It was decided to invite him all over again, and accord his state the privileges in the new union which his brave stand had merited. He accepted two weeks later.

Sir William Luce and the British government were well content. What did it matter that the windows of some British offices were stoned, and that mobs in Kuwait and Bahrain burned union jacks? The rulers understood what had happened, even if the mobs did not. And the object of the exercise had been achieved. Iran had the all-important Islands of Tunb, and Ras al Khaimah was a member of the federation.

In fact, as it turned out, there was one leading Arab who had not in any way 'understood' what had happened over the occupation of the Tunbs, and that was Mouammar al Gaddafi of Libya. He did not appear to be aware that the mild denunciations which were now issuing from government offices in Cairo, Kuwait, Baghdad and Riyadh came from regimes which had absolutely no intention of doing anything about the situation, and their embarrassment was considerable when Gaddafi called them to band together and throw the Iranian aggressor back into the sea. Most embarrassed of all were the rulers of the new Union of Arab Emirates, especially when Gaddafi offered them direct aid in what he presumed would be their imminent counter-attack against the invader.

It so happened that on 5 December 1971, six days after Iran's occupation of the islands, OPEC met in Abu Dhabi for one of the organization's regular conferences. Libya had announced that it was boycotting the meeting because it refused to sit down at the same table as the Iranians. But OPEC's discussions were to be followed by a meeting of OAPEC (the Organization of Arabian Petroleum Exporting Countries)* and to this the Libyans had sent their usual delegation. While OPEC settled down to the discussion of its agenda, the Libyans took off for Ras al Khaimah, some hundred and eighty miles along the shores of the Gulf. There they burst in upon the astonished Sheikh Sakr and, to his considerable confusion, passed on to him a message from Gaddafi offering him Libyan troops and arms with which to fight back against the Iranians. The ruler, though he had no intention of taking it seriously, thanked the delegates most warmly for their offer, and gave them a signed photograph of himself to take back to their leader.

Back in Abu Dhabi, the Libyans informed the oil ministers of Abu Dhabi and Dubai of the nature of their visit to Ras al Khaimah and told them that, in the event of Libyan troops being sent, passage would be needed for them through their territories. The ministers promised to convey this information to their rulers, who must have been considerably more alarmed at the prospect of having Libyan troops in Abu Dhabi and Dubai than Iranian forces on the Islands of Tunb. Fortunately for them, Sheikh Sakr let the Libyans know that he could rely upon his Arab brothers in the Gulf for any help he required, and that Libyan troops would not be needed.

Frustrated in his hopes of hitting back at the Iranians, Gaddafi turned upon the perfidious British, who had been responsible for the whole thing, anyway. Against the British, at least, he could retaliate without having to ask anyone else's by-your-leave.

On 7 December 1971, he took over Britain's oil installations in Libya and announced the nationalization of British Petroleum.

'He's closing us down!' said a BP engineer in Tripoli, when news reached head office in Libya of Gaddafi's action. 'I wonder if he knows what he's doing?'

'Let's not tell him, anyway,' said one of his managerial colleagues.

The crude oil which comes out of the Libyan desert has qualities which make it particularly desirable on the markets

* OAPEC was formed by the Arab oil states in 1967, after the Six-Day War.

of the world. It has an extremely low sulphur content, and is particularly valuable for gasification, for producing chemical feeding stocks, and for high-grade petroleum.* But it has one element in it which can cause problems and that is the amount of wax in the oil. Petroleum engineers use mechanical gadgets called 'go devils' to keep the oil from clogging the pipelines on its way from the wells to the loading ports, and they are particularly vital in the Libyan fields because the waxy content of the crude can clog the pipes in the space of hours.

For two or three days after the seizure of their assets in Libya, followed by the withdrawal of personnel and the closing down of operations in the field, BP officials hugged to themselves the thought of what was going to happen to Concession 65 if some action wasn't taken quickly. The wells themselves would begin to clog up. The crude in the tank-farms would begin to coagulate. As for the pipeline from the field to Tobruk, the loading port, it would very soon become what one engineer described as 'a bloody great wax candle, fit for nothing except chopping up into sections and putting on your dinner table'.

In the circumstances, some of the more irresponsible BP staff can perhaps be forgiven for the quiet pleasure this prospect gave them.

One of those who did not share any delight in the imminent seize-up of Concession 65 was an American named Nelson Bunker Hunt. Bunker Hunt is one of the three sons of the Texas oil billionaire, H. L. Hunt, and he and his brother, Lamar, owned and operated their own oil company, Hunt International Petroleum Company. Hunt International was a part-owner with BP of Concession 65, and the Americans had had their assets seized by the Libyans at the same time as those of their British partner. Bunker Hunt was furious. He flew in from Texas to Tripoli to protest to Gaddafi's right-hand man, Major Jalloud, pointing out that he was not British, he had no connections with Iran, and no operations in the Gulf. Why hold him up to ransom? Hadn't he always been a good friend of the Arabs?

When this demarche failed to have any effect, Hunt took off for Algiers, where he and his brother also had operations, and from the Algerian premier, Boumedienne, he got both

---

* Oil from the Gulf, for instance, is apt to be high in sulphur, resulting in pollution problems at the refineries and in tankers.

sympathy and help. Boumedienne at once contacted Gaddafi to point out Hunt's friendly attitude and his 'unswerving sympathy for the Arab cause', and urged him to keep Hunt apart from any quarrel he might have with the British.

The appeal was immediately effective and the seizure of Hunt International's holdings was cancelled. Gaddafi's gesture could not have been more fortunate for himself and for Libya. For, the moment he was reinstated, Bunker Hunt looked over Concession 65 and warned that unless something was done – and done quickly – the oil would solidify in the pipes and something like $10,000,000 worth of damage would be done to one of the most prosperous oilfields in the Western Desert.

Within twenty-four hours he had been given permission to fly in thirty-five of his own experts from Texas, and under their supervision the field was got back into operation. Concession 65 was saved.

It is doubtful whether the news of Bunker Hunt's last-minute intervention raised any cheers in Britannic House, BP's headquarters in the City of London. The British company, which knew how desperately short Libya was of first-class oil technicians, had counted on this dearth of personnel to force the Libyans to the negotiating table. Now this hope had been dashed by their American partner, and their £100,000,000 investment in Libya was back in operation – but out of their hands.

It was just like 1951 all over again, when Mossadeq had nationalized their oil operations in Iran. Only this time they could not hope that the US major oil companies would come to their rescue. The climate had changed, and so had the nature of the oil industry. There were too many independents around, for one thing, for the majors to enforce a boycott. And, in any case, why should they? In Iran, by helping BP, they had gained a sizeable stake in Iranian oil. But in Libya, they already had their stakes. So what could they gain by helping the British?

So BP had to content themselves with protesting the seizure to the Libyan government, condemning it as a clear violation of international law, and calling for arbitration of the dispute between the company and the government. In the meantime, the company took space in the world's newspapers warning any would-be purchasers of oil from Concession 65 that they would be buying 'pirate' oil.

The company has also reminded the Libyan Government [BP's statement concluded] that the government's wrongful acts are under international law incapable of depriving the company of its rights under the agreement. Accordingly, the attention of all those who may be concerned with these developments, whether as purchasers of oil or otherwise, is drawn to the continuance of the company's rights. It is the intention of the company to assert those rights wherever and whenever necessary against those who infringe them.[5]

It was the same action as BP had taken during the dispute with Mossadeq, and, as in the case of Iran, the company planned to prosecute anyone anywhere in the world who tried to buy or sell oil from Concession 65.

In 1951, the threat had worked and no one in the world had dared to buy. Iran had been starved into submission. But this was twenty years later, and unlike Iran, Libya was one of the richest countries in the world, with something over £3,000,000,000 lying around in the bank, ready to be used in just such an emergency.

PART EIGHT

Crude Facts

26 · Private Club

In February 1972, members of the Organization of Petroleum Exporting Countries (OPEC) met the major Western oil companies in a series of confrontations in Teheran, and by bluff, threats and smart manoeuvring squeezed giant increases in their revenues out of the companies. It was probably the most expensive commercial negotiation ever, and would, by 1975, cost the major oil combines the astronomical sum of £15,000 million in extra payments to the Arabs and Iranians.

Kuwait alone had its 1972 income from oil doubled from £300 million to £600 million, which is not bad going for a state only 10,000 square miles in size and with a population of around 700,000.

There is a story among oilmen in Kuwait that five times a day, when he kneels down to pray, the ruler, Sheikh Sabah al Salem al Sabah, inevitably ends by saying:

'Allah, tell me and my people what we are to do with all our money?'

With the possible exception of Abu Dhabi, which has an even smaller population with which to share its oil wealth, Kuwait is the richest state in the world, and it shows. The tiny pearl- and prawn-fishing village of thirty years ago is now a land of broad highways, modern hotels and fantastic villas, of shops stuffed with expensive imported clothes, food, jewellery, watches and perfumes. It has fourteen hospitals with a bed for every eighty Kuwaitis, and the medical equipment is the envy of the medical world. It has 230 public and 40 private schools, and a teacher for every ten pupils. Lessons are aided by the latest audio-visual aids, leisure is encouraged by tennis courts, swimming-pools and games arenas in every school – all free.

All the facts about Kuwait are fabulous. One in three Kuwaiti citizens now owns a car, and that statistic includes babies and women, only a few of whom drive. The cars are mostly big and American, and Kuwaitis vie with the Japanese as the most dangerous drivers in the world. In the desert sands alongside the highways leading out of the city, brand new Cadillacs and Chevrolets and Plymouths lie like crushed beer cans, crashed at break-neck speed, seized-up for lack of oil in the sump, or merely dumped because of a blown-out tyre. Their owners prefer to buy new one rather than bring them in.

Land in the centre of Kuwait is now £180 a square foot, and yet hotels and office buildings continue to rise, and so do new schools, parliament buildings, exhibition halls and ministries. The state has built the biggest desalination plant in the world, and water is used lavishly, for gardens and small parks, to quench the thirst of the oleanders which line most of the highways, and this in a land which has one of the hottest summers in the world (it averages 115 degrees Fahrenheit in the shade in August). Yet Kuwaitis do not mind the expense because it is not they but oil revenues which pay for it. For them the telephone is free; they are flown by the state to New York, London or Paris if they need special medical attention; they pay no taxes on their incomes or houses, basic rates for electricity and gas in their homes, and petroleum in their cars; and any Kuwaiti can go to the government and borrow 25,000 dinars (£30,000) free of interest to build himself a home, and pay it back at a rate of not more than five per cent of his salary over twenty-five or thirty years. These children of the desert have constructed air-conditioned villas on the sands where they once grazed their camels and lived in goatskin tents, and on summer evenings they come out of them, trailing TV sets and fans and refrigerators on long extension wires, and they squat in the sands as of yore, sipping icecold Pepsi Cola and watching 'Bonanza' or the latest film from Cairo.

But these incredible benefits of a super welfare state are the privilege of Kuwaiti citizens only, and it is not easy to become a member of the club. Of the state's 700,000 inhabitants, less than 400,000 are Kuwaitis. The rest are Palestinians (about 100,000), Egyptians, Iranians, Iraqis, Jordanians and Saudis who have come swarming into the desert Eldorado in search of work. They have no trouble in finding it, for why should a Kuwaiti work when he has all that money? But jobs do not bring with them the privileges that a citizen automatically enjoys, and only those who have some special skill to give the state are granted citizenship. The rest do most of the work that keeps Kuwait going, and watch their rich masters taking it easy.

Out in the desert, just beyond the confines of the city, are the bidonvilles of the non-Kuwaitis who will never have the special skills that would qualify them for citizenship, and who have come from all points of the Middle East to wait for the crumbs dropping from the rich man's table. They live in an atmosphere laden with the stench of burning oil, and their squalid hutments are lit at night by the flares of burning gas. They have no air conditioning to cushion them against the appalling heat of summer, and little protection from the violent sandstorms for which Kuwait is notorious. In winter they are whipped by storms and their settlements turned into brown seas of mud by torrential rain. They are fellow Arabs or Iranians who, by an accident of geography, come from outside the confines of Kuwait and therefore do not qualify for a share of the riches, and perhaps they would not resent their misfortune so much if the Kuwaitis were not so contemptuous of them.

The Kuwait government pays some 60,000,000 dinars (about £70,000,000) a year subvention to Egypt and Jordan, and is prompt in paying its dues to the Arab League, but in individual encounters with brother Muslims who have come to work for him the average Kuwaiti is apt to treat him like a servant.

'It is bad enough to see the young men doing no work, flinging their money around, wasting it on useless extravagances,' said one Palestinian, bitterly, 'but even worse is their arrogance. I always thought that one of the faults of us Palestinians was our superior manner, but compared with Kuwaitis we are humble people. They will not allow us to buy land, or go into business without a Kuwaiti partner, or be medically attended without paying a fee – but all that we could stomach if only they would acknowledge the contribution we make to the running of their state. Without us it would stop functioning. They go to college but come away with easy arts or sociology degrees, and they wouldn't know a bolt from a spanner. So they sit around in offices where the government has invented jobs for them, and they pass out nonsensical orders that we take no notice of. If we did, everything would go wrong! And they have the impudence to look down on us – just because they have money!'[1]

One of these days, some prophets darkly forecast, a solution will be found to the Israel-Arab situation, and then the Palestinians in Kuwait will need a new cause to which to direct their passions. It could well be a campaign to force a more equitable share of Kuwait's riches for those who keep the wheels of the state turning. It is perhaps the vision of this un-happy day, and the knowledge that they are envied (and in some cases, hated) by most of their neighbours which makes Kuwaiti males the febrile, haughty, unstable characters that they are. Oil wealth seems to have brought them only uneasiness, suspicion, and xenophobia.

A *New York Times* reporter surveying the Kuwaiti scene in 1971 quoted a psychiatrist's observation that outpatients in the city's mental hospitals had risen from 450 in 1956 to 18,000. The psychiatrist added:

Many of [their] neuroses are related to the way the Kuwaiti has come to consider himself an elite person. It is not important in his eyes whether he works or not. If he does something wrong he feels it can be excused. He looks down on naturalized citizens as second class and non-Kuwaitis as quite something else again. These people come to me and say they can't sleep, can't work and can't enjoy life.

Some of Kuwait's leaders [the writer added] have begun to feel that the perpetual onward and upward financial spiral has brought no rewards in terms of happiness and they are calling for a slow-down, a time for Kuwaitis to catch up emotionally and culturally with their new economic status.[2]

It does not help the stability of the Kuwaitis that they find themselves constantly looking over their shoulders at their neighbours, by whom they are not only regarded as the spoiled brats of the Persian Gulf but against whom all of them have territorial claims. The vague frontier agreements which were made by Sir Percy Cox at Oqair in 1923 satisfy neither the Saudi Arabians nor the Iraqis, and Iran has serious objections to the extent of Kuwaiti territorial control over the off-shore regions of this cul-de-sac area of the Gulf. They are three nations whose considerable oil revenues fall far short of their budgets, and they can be forgiven for the baleful looks they cast at a city-state which is quite often perplexed about what to do with its money.

Twenty-five miles south of the city of Kuwait is Ahmadi, head-quarters of the Kuwait Oil Company, an Anglo-American con-dominium whose 692 wells keep half the world's industrial complexes supplied with the power to keep them going. When people in the Western world talk glibly about the 'rapacious' or 'capitalistic exploiting' Anglo-American oil industries of the Middle East, they can have no conception of places like Ahmadi in Kuwait, Dhahran and Al Khobar in Saudi Arabia, and several company towns in Iran where millions of dollars have been spent on the people.

Ahmadi is a company town, and all of its 30,000 population are in one way or another connected with the operations of the Kuwait Oil Company. There are, as in Kuwait city, houses for them to live in, swimming-pools and tennis courts and recreation grounds where they can play, theatres, cinemas, and yet another first-class hospital with top surgical experts attached. But at least here there is no discrimination against the non-Kuwaitis, and the 300 British, 44 Americans, 1,400 Kuwaitis, 280 Indians, 160 Pakistanis, and 1,600 non-Kuwaiti Arabs all have the same privileges, and the same attention. So do their appendages, as well as the Kuwaiti members of the fire service, police, postal services and maintenance facilities who help to keep the town going. They are a financial charge on the oil company, and it comes out of their financial take.

It is a green town where British housewives have encouraged their Kuwaiti and Indian opposite numbers to cultivate their gardens, and massive amounts of desalinated water go into the egging on of grass and flowers in the gardens of company houses. In March and April they are a sight to see, green oases against which riots of colourful flowers bloom, only to die in the onslaught of heat at the beginning of May. And each week-end cars stream out from Kuwait packed with families come to see the soothing green and picnic on the streets. And when the police try to move them on, they refuse to budge. 'It is our town,' they say.

As, indeed, it is, though it is paid for and administered by KOC. And sometimes, a weary American or British admin-istrator of the town, squeezing municipal affairs in between his normal work, expresses a hope that one day the Kuwaitis will take over the place. But 'each time we suggest it, they shrug their shoulders and say as long as it runs well, why should they bother. I wish they would take the same view of the oil operations.'[3]

Ten miles out into the shallow waters of the Gulf, a 66-inch undersea pipeline – wide enough to take a Volkswagen or a Mini – pumps 30,000 barrels of crude an hour into the bowels of 250,000 ton tankers riding alongside Sea Island, KOC's feeding point for its fecund fields. But sometimes storms blow up in the Gulf, and a great wind rides across the desert, mixing sand and oil-smoke into a choking fog. The empty tankers are forced to take their vulnerable balloon shapes out into deeper waters, where they cannot be blown ashore; and then the marine masters watch the turbulence of the sea and all the other weather signs, and calculate how long they dare wait before calling the tankers back in to link up with the pipelines. If they wait too long, the tank-farms fill up and threaten to slop over with surplus oil, and signals must go out to the feeding points and well heads, closing them down.

No oil company official is more alarmed than the rulers of Kuwait when this situation threatens. In the spring of 1972 a storm blew up which raged along the shores of the Gulf for nearly eight days, forcing tankers to put back to sea, feeding operations to stop, and oilwells to be closed down. No one informed the population of Kuwait of what was happening, because the volatile population of the city might have panicked had they realized the situation.

For Kuwait still runs on gas from the oilfields and the state would grind to a standstill without it.

'A state like Kuwait,' Ahmad Zaki Yamani of Saudi Arabia once said, 'could not survive without the oil – or rather the gas produced in association with oil – from which is generated the power to produce its water and electricity.'[4]

In the storm of 1972, the city came within four hours of closing down as the oilwells cut out and the gas tanks began to empty. And after the storm had blown itself out, the wells had started up again, and the tankers had come back alongside, a KOC official said:

'In a way, it's a comfort for us to know that this can happen.

L

And for them to know too. When the Kuwaitis are feeling particularly bloody-minded with us, usually because of some action totally apart from our operations, something to do with American or British policy towards Israel or Vietnam or something like that, their militants start threatening us. "We'll show you who's master here," they say. "With one turn of a spigot, we can deny the West the oil it needs from us. We will call a general strike and close down the oilfields." At which, the government in Kuwait rushes out, wringing its hands, and says, "For God's sake, no. Don't you realize what would happen? No gas. No power to give us light or air conditioning. No power to pump our water. We will die of heat prostration, or, if we survive that, of disease." It is one thing we can be sure of. They can starve us, beat us, shoot us, make life unpleasant in the extreme. But, for their own sakes, they can never close us down.'[5]

27 · Participation

On a crisp January afternoon in 1972, I drove through Riyadh, the Saudi Arabian capital, to a meeting with Ahmad Zaki Yamani, the king's minister of petroleum and mineral resources. We were going to talk about the future of foreign oil companies in the Middle East, but at that moment nothing could have seemed more remote.

It was getting on towards the end of a bitterly cold day of the kind that always surprises the visitor to Central Arabia and makes him wish he had brought warmer clothing. The sun was setting and the sky, which had been a brilliant blue all day, was now turning crimson, its glow washing over the walls of the palaces, ministries and mosques which line the broad streets of the city's modern quarter. Everything was stained red by the sun, even the black, round, cheerful face of my bedouin driver; and reddest of all were the uneven crags and shallow defiles of the desert, rising and falling towards the horizon at the end of every street.

From several mosques within earshot the loudhailers in the minarets were calling the faithful to the fifth and final prayer of the day. In the time of Ibn Saud, everything would have stopped at the call of the muezzin, but now, though groups of men hanging around the government offices began to fall to their knees, the traffic went on rolling down the streets – big American cars, lots of heavily-laden trucks, an occasional bus, and not a camel or donkey in sight.

Riyadh is capital of the richest oil state in the world, but doesn't look like it. It is a small, clean, sandstone town dominated, of all things, by a giant water-tower whose slender stem and mushroom cap catch every glint of light. It is floodlit by night and visible for miles around. Some see in its shape a symbol of Saudi Arabian virility, but to the traveller glimpsing its rearing head from across the desert the earthly delights it suggests are not likely to be fulfilled when he reaches the city. When the US boxer, Mohammad Ali, was staying there as a government guest, he was heard to remark: 'Doesn't anyone ever sing or smile in this town?'

The answer is that only the muezzins sing, and there is really not very much to smile about. Riyadh has always had the reputation of being Arabia's most inhospitable city ('saturnine and sanctimonious' was what Ameen Rihani called it), and even Saudis from other parts of the country find it unbending and unfriendly. It is not simply that there are no cinemas, theatres or other places of entertainment (they are forbidden in other parts of the country too), but that there is a pervading air of grey sanctity about the place, a nip of prudishness in the air, which strangers find inhibiting and seems to make the natives glum. Arabs who know both places say that Mecca is much more cheerful. As for Western infidels, they live not in Riyadh but on either side on Saudi Arabia's two coasts, in Dhahran or Al Khobar on the Persian Gulf if they are in oil, in Jiddah on the Red Sea if they are in shipping or trade,* and they make short visits to the capital only when vitally necessary.

Ahmad Zaki Yamani, Saudi Arabia's minister of oil and mineral resources, is an unlikely figure to find thriving in such a restrictive atmosphere, for in manner, appearance and outlook he gives every sign of wide and worldly sophistication. Of all the oil ministers with whom Western negotiators have to deal, Yamani is the one with whom they most seem to 'get along'. Howard Page, who faced him several times across the table on behalf of Standard of New Jersey, said of him:

I know damn well that there are occasions when Yamani will insist on doing something which he knows is not in his country's best interests, but that's because he has got himself hooked politically.[1]

This ability to persuade Western oilmen that he is doing his best to meet them, though under pressure, while at the same time convincing his militant Arab and Iranian colleagues that he is squeezing the last drop of blood out of the companies is Yamani's principal asset as a negotiator. But he has others.

* Or in diplomacy. All foreign embassies are in Jiddah, as is the Saudi Ministry of Foreign Affairs. Envoys come to Riyadh only when summoned there by the king.

He's very clever [said Page]. I remember when we were working on negotiations with OPEC and we had come down to Saudi Arabia. Things weren't going well, and Yamani needed an inspiration to get the negotiations unclogged. He took me and the other Western delegates to see King Faisal. The parent companies of Aramco were arguing about some tax situation with the Saudis. Well, Yamani got me right up in front when we got to the palace, closer to the King than you are to me, and it turned out that he had Faisal well briefed. The King hauled me over the coals for not agreeing to some of the proposals about taxes that OPEC had made. The hell of it was that I was perfectly willing to agree to the proposals, but the two other fellows squatting there on the cushions (from two of the parent companies) were not. Of course I couldn't say so. But I think Yamani knew the score, and what he was trying to do was break me down so that I would say to Faisal: 'I agree. It's only these two fellers who are holding it up.' To put them in a spot and make them back down.[2]

For his private home in Riyadh, the Saudi oil minister has annexed a wing of the capital's newest hotel and surrounded it with a garden lush with green grass, palms, oleanders and hibiscus. Inside, concealed lights glow on vivid silk carpets hung on the woollen-padded walls. Underfoot the carpets are deep and soft and strewn with low-slung damask chairs. A hi-fi system plays tapes from Western musicals and films, so that a conversation about oil or Arab politics is apt to be carried on against a background of the theme music from *Doctor Zhivago*. Sometimes there is a rustle of silk through the filigreed divisions at the end of the room and the faint scent of French perfume, and one remembers the smart, good-looking woman who is sometimes seen with Yamani on his trips to Beirut, Teheran and San Francisco, modern and unveiled on these occasions, but relegated to her own quarters in the restricted conditions of the capital.

Yamani has a nightmare which, he fears, may yet come true. It is that one day all the Middle East oil states, in a fit of nationalistic rage, probably as a result of an exacerbation of the Israeli-Arab situation, will turn on the Western oil companies operating inside their countries and nationalize the lot of them, just as Algeria has done to the French, and Libya and Iraq to the British. It is a policy of annexation preached by all local militants ('Arab oil belongs to the Arab people'), and it could, Yamani believes, do nothing but harm to the oil states, economically, sociologically and politically. He does not want to see the major oil companies booted out of his own or any other country in the area, and his reasons are selfishly pro-Arab.

When he talks about it, three terms come into his conversation, as they do into the talk of most oilmen, and it might be as well to deal with them here. The first is 'upstream'. *Upstream* means all the activities of the oil companies in the country where the crude is actually pumped out of the wells. The second is 'downstream'. *Downstream* is the outside world, where the oil is actually marketed. The third is 'off-takers'. *Off-takers* are companies which have no part in the upstream production of oil, but only in the downstream business of selling it.

Under the present set-up [Yamani once said] the majors make the bulk of their profits in the producing end of the business. In a country like Saudi Arabia, for instance, they produce the oil at a cost of, say, 15 cents a barrel, perhaps less; and they pay Saudi Arabia perhaps 90 cents a barrel in royalty and income tax. Now Arabian light crude, for example, has a posted price of $1.80 per barrel; but its realized price in the market may be only $1.40.* So if the majors then feed this oil into their refineries at the posted price, this means that they do not actually make any profit in their downstream operations, and in some markets may even register losses. Thus virtually all their profits . . . are made upstream at the production stage.[3]

For this reason, Yamani points out, no major oil company under the conditions under which it operates today is really interested in bringing down the price of oil, either crude or refined. On the contrary, the majors have a vested interest in keeping up the price, and, 'one might say that they are now really the only available bulwark of stability in the world market'. The producing countries should welcome this situation, because, under the existing set-up, they can count on a regular and stable income, enabling them to plan ahead and fix their budgets.

Now, against this background, let us look at nationalization [says Yamani] and see what will happen if we nationalize the producing operations in our countries . . . Nationalization of their upstream operations would inevitably deprive the majors of any further

* The figures given are for 1969. The posted price is, of course, now very much higher, reinforcing Yamani's point.

interest in maintaining crude oil price levels. They would then become mere off-takers buying the crude oil from the producing countries and moving it to their markets in Europe, Japan and the rest of the world. In other words, their present integrated profit structure, whereby the bulk of their profits are concentrated in the producing end, would be totally transformed. With the elimination of their present profit margin of, say, 40 cents a barrel from production operations, the majors would have to make this up by shifting their profit-focus downstream to their refining and product-marketing operations. Consequently, their interest would be identical with that of the consumers—namely, to buy crude oil at the cheapest possible price. They would put their full weight behind efforts to drive down crude oil prices, and in this they would undoubtedly succeed.[4]

There could be only one result from such a situation, Yamani maintains:

A dramatic collapse of the price-structure, with each of the producing countries trying to maintain its budgeted income requirements in the face of the declining prices by moving larger volumes of oil to the market. I need hardly emphasize the disastrous effects of such an eventuality on the economic and political life of the producing countries. Financial instability would inevitably lead to political instability.[5]

He admits that the consuming countries might, for a time, enjoy a glorious period of cheap oil at their power plants and gas stations, but believes that the chaotic conditions which would arise in the Middle East would, in the end, outweigh any economic advantage which might result from the change.

The chief apostle of nationalization of the Middle East oil-fields is, in fact, Yamani's predecessor as Saudi oil minister and his bitter rival, Abdullah Tariki. Tariki operates today as one of that new breed of experts, a petroleum adviser, and in that capacity he gives advice on tactics and strategy to the oil administrations in Algeria, Libya, Iraq and Kuwait. In the first three countries he has succeeded in persuading his clients to seize all or some of the oil operations conducted there by the major companies, but no one in the oil world is in any doubt that Tariki's main target for nationalization operates inside a country for which he has no mandate whatsoever to act as an adviser. In other words, Aramco of Saudi Arabia. He has always been antipathetic towards the great American combine. As for Saudi Arabia, though it is his native land he has a consuming hatred for the present administration, which has driven him into exile, which is trying to persuade friendly governments to ban him entirely from the Middle East, and whose ruler is his implacable enemy.

Tariki was in Baghdad when the Iraqi government nationalized the northern territory operations of the Iraq Petroleum Company on 1 June 1972, and those who listened to President Bakr, leader of the Iraq Revolutionary Command Council, announcing the takeover guessed that the Saudi adviser had not only master-minded the act but also written the words to go with it. They exactly echoed his own fiery sentiments.

You know [announced President Bakr] that the oil companies are the dangerous tools which represent imperialist logic, the logic of plunder and monopolistic exploitation and the impoverishment of the masses. They have always continued to represent imperialist domination, both symbolically and in fact, so that it has become clear and obvious that any true national liberation is incomplete unless the requirements of national sovereignty are imposed on these companies which have acted in accordance with their imperialist nature, contrary to the interests of our masses, the spirit of the times and the progress of history, and have considered themselves a state within a state. It is clear to the struggling masses from their experiences that putting an end to the domination of the monopolistic oil companies is the only way to ensure national sovereignty and the economic independence which is the tangible essence of political independence.[6]

It was genuine Tariki rhetoric, full of the references he loves to make to imperialism and plentifully sown with his favourite trigger words, like 'struggling masses' and 'economic independence' and 'national liberation'. He was not allowed to proceed to Beirut for the emergency meeting of OPEC which was called, on 9–10 June, to discuss the Iraqi nationalization, for the Lebanese government, at the request of the Saudis, has declared him persona non grata. But his clients among the delegates came well-primed with his arguments, and rumours in the OPEC lobbies had it that the Iraqis had been told by Tariki to rest assured that within a year of IPC's nationalization that of Aramco would follow. After that, the rest of the oil concessions of the Middle East would automatically be 'liberated'.

It was a difficult OPEC conference for Ahmad Zaki Yamani, for it eventually endorsed the Iraqi nationalization of IPC and promised the full support of the oil-producing nations to its fellow member in the event of any trouble with the Anglo-American-French-Dutch directorate of the seized company. By voting to help make Iraq's annexation a success, Yamani was not only aiding an enemy of himself and his state, but he was also sabotaging his own plans for solving the crisis which faces oil states and oil companies in the 1970s.

For Yamani's objections to nationalization are not simply negative. He has been the most active supporter among the ministers of the oil states for what he believes is the more sensible alternative to annexation, which is participation by the producing countries in the operations of the major companies. Ever since Enrico Mattei started the trend in Iran in 1958, new concessions handed out by Middle East oil states have had a clause written into the contracts stipulating that, once petroleum was discovered, the concessionaire would have the host country as an active partner in the operations. Ahmad Zaki Yamani, for example, is now an active director of the Arabian Oil Company, the Japanese combine which has discovered rich oilfields offshore from Saudi Arabia and Kuwait, and he participates in all the higher policy decisions which are taken by the company.

So far as the major oil companies are concerned, however, the concessions are of the old type (even those which have been revised) operated by foreigners whose principal responsibility is to their parent companies or shareholders in Europe or the United States, and in whose plans and policies the host countries have no knowledge or say whatsoever.

Let me take Aramco, for example [says Yamani]. Since Aramco is incorporated in this country [Saudi Arabia] I, as Oil Minister, have been made a member of the board. But that's just show business. It is meaningless. Aramco has what it calls its executive committee, and that is where they plan the real business, where they get instructions from their parent companies in America. *I want to be there.* [He speaks with great emphasis.] I want to know what is going on and I want to have a say in it. And that is what frightens them most.[7]

Yamani has been trying to make a participation deal with

Aramco ever since 1963, and has been fobbed off repeatedly by a succession of US presidents of the company.

It isn't the financial aspects of participation which frighten the company—and all the other companies—most, though they insist that this is the question that bugs them most [he says]. What they are worried about is having Arabs or Iranians actively taking part in the direction of the company. They are afraid that we might not have the know-how, we might over-estimate our ability, or we might interfere. Take Aramco again. I wanted them to relax on this point, so in 1963 I started asking Aramco to go into a joint venture with Saudi Arabia on a small scale. They said: 'Why?' I said: 'Just to know each other, nothing more. To do business together.' We even found a third party to be with us to make it easier for them. They refused.[8]

In 1968 he tried again.

There is an oil structure lying partly inside Aramco's concession and partly in the Saudi government's own concession. I came to them and I said: 'How about developing this oil structure together? If we find oil, we can go into it together.' They couldn't say No by this time, because by 1968 foreign companies couldn't afford to be rude to host governments any longer. So they said: 'All right,' and they went off to America, obviously to consult with their parent companies. Then they came back to us, not saying no again, but you know how people go on when they don't want to do something but don't want to refuse outright, having an argument, playing for time.Then we announced an invitation to any comers to go into our area with us, and when we did that, they came running to us, and said: 'Let's do it now.' But then it was too late.[9]

By 1971 the oil states of the Middle East were beginning to divide up into militants and moderates, and so far as Ahmad Zaki Yamani was concerned, Tariki had become the militant thorn in his moderate flesh. All around him he could see signs that Tariki's activist policy of nationalization and takeover was gaining ground, with Arab leaders in some of the countries, and with the Arab and Iranian masses in all of them. Apart from the National Iranian Oil Company – which had special relations with the Anglo-American consortium – there was not a single national oil company in the Middle East which was running a successful independent operation. The prime example of abject failure was in that phenomenally rich little oil state, Kuwait, where conditions were such that it should have been impossible to make a loss out of oil. Yet the local national oil company had succeeded in doing just that. Its brand new refinery had failed to cope adequately with the free oil handed over to it by the Kuwait Oil Company, and its marketing plans had gone so sadly awry that Russia was buying up Kuwait National's oil cheap for sale to its clients in Asia and selling to Kuwait National in Europe refined oil at high prices for its markets there, entailing losses all round. In Saudi Arabia, too, the national company was having trouble with its petrochemical division at home and its marketing abroad; and almost exactly the same situation prevailed in Iraq, where the Iraq National Oil Company (INOC) had had a particularly bad year.

In the circumstances, it was not difficult to convince reasonable men that nationalization of the major oil concessions was the worst possible solution – at least for a decade – to the problem of the Middle East's control of its own oil. For Yamani and his fellow moderates, however, the difficulty was that the Iranian and Arabian masses were not reasonable men, because militants like Tariki and the masters whom he served (like Gaddafi of Libya, Bakr of Iraq and Boumedienne of Algeria) insisted on propagandizing and politicizing the oil situation, picturing the major oil companies as the enemy whom they could defeat only by seizing their concessions.

'Lord Balfour once said that, pity though it is, most colonial people prefer self-government to good government,' said a British oilman in Iran. 'It's the same with these people. It's foolish of them, but the Iranians and Arabs prefer to run their own oilfields badly than have them run well by outsiders, particularly British and Americans.'

Yamani's task was to try to convince the governments of the oil states (he knew he had no chance of convincing the masses, whipped up by Tariki-type rhetoric) that participation was the better way, because it would achieve control by gradual means without interrupting the flow of oil upstream or its profitable sale at a stable price downstream. His arguments, his passionate sincerity and the wealth of documentation which he produced had so convinced Middle East leaders that by mid–1971 he had the backing of all member-states in OPEC for a joint demand for participation. The producer states would approach the major Western concessionaires and ask for an immediate twenty per cent share in their operations, this share to be increased as time went on and the Middle East gained more technical know-how.

Tariki fought hard to persuade his client-states to veto the proposals, but failed. Libya, Algeria and Iraq indicated that they would make their own policies, but otherwise went along with the mood of the seven other members.

But now the task was to convince companies and combines like Aramco and BP and Standard of New Jersey and Shell and Gulf and Mobil that they must now make a gesture to the moderates and sacrifice some of the sovereignty of their concessions to the states in which they were operating. Yamani was well aware that this was going to be the hardest part of his struggle, and that Western oilmen didn't believe a word of it when he insisted that the sacrifice would be for their own good.

Unfortunately, the majors—at least some of them in their public statements [he had once said] seem to be obsessed with the empire they have built. It is so vast and it took them so many decades to achieve. And now they see these newcomers—these national oil companies in the producing countries—wanting to come and take a piece of their cake, which is the last thing they want to happen.[10]

In a later part of the same speech, he made an indirect appeal for cooperation from the majors:

Participation serves so many purposes [he said]. First and foremost it will maintain the stability of prices. It has some common features with nationalization, but without the drawbacks. From our point of view, it will give us [the oil states] a position of influence in the market, instead of being pawns in the hands of other people. It will be a good thing for the oil companies because it will save them from nationalization and provide them with an enduring link with the producing countries. The majors' concessions will be starting to expire in 20 years or so and, unless they have this bond with the producers, their very existence will be in jeopardy later on. They will have nothing at the producing end and their position in the market will be weak.[11]

The reaction of Aramco at that stage was blunt.

'Aramco did not have a negative attitude towards participation.' Yamani said later. 'They simply said no.' He added, crisply: 'But we will make them say yes.'

By the beginning of 1972, anyone travelling around the Middle East oil states could not fail to be aware that the crunch was coming. Algeria had already taken over the Sahara oilfields from France. Libya had nationalized BP and there were strong rumours that Standard of New Jersey's concessions were next on the list. In Iraq the Baathist government was increasing its pressure on IPC. In Iran the shah had made several statements attacking the 'trickery' and 'unfair economic pressure' which had been used by the Anglo-American consortium to make his people give in to their demands in 1954, and he had gone on to threaten reprisals when the concession came to an end in 1979.

In Saudi Arabia, Ahmad Zaki Yamani was worried. He had rallied the moderate OPEC states behind him against nationalization and in favour of participation, but no gesture had come from the major oil companies. And time was running out. The militants were gaining ground.

If I were a consumer in the West [Yamani said] and I looked down the road to 1980, I would feel rather scared that some time in the eighties I would not find enough energy for my industry. I would also be worrying about the nationalist trend which is prevailing now all over the world, mainly in the evolving countries—the national pride they have vis-à-vis their natural resources. The only way I would feel secure about my sources of oil supply would be if I had accepted the principle of participation, and my interests were tied up with those of the producing countries.[19]

Yamani was speaking (it was 3 January 1972) almost immediately after a meeting with Aramco officials, and he was evidently depressed by the way his discussions had gone. In two weeks' time, OPEC would be meeting the majors in Geneva to force a decision out of them over participation, and one gathered that he had failed to impress upon the American directors of Aramco, in these preliminary talks, that the situation was approaching a crisis.

The trouble with the Americans is that they don't look at it the way we do or the consumers do in the non-American world [he said]. They think they have their own reserves. They have Alaska. True, already the United States imports half of the oil they consume, but they have Venezuela next door, and they think right now that with the help of their major oil companies they can satisfy their needs. But I think they are wrong. I think that by 1979 there will be a drastic change in the oil industry, especially here in the Middle East. The Iranians have made it very clear that they will not live with the Anglo-American consortium after that date. Then, even though the concessions are supposed to last until the year 2000 and longer, if the oil companies think that they can live as they are with their agreements in Kuwait, Saudi Arabia, Abu Dhabi and elsewhere, then I think they are fooling themselves. If Libya goes ahead and nationalizes all its oil, if Iraq follows suit, then responsible ministers in places like Kuwait and Saudi Arabia will have to take cognisance of Arab pride, and we won't be able to keep quiet, or remain moderate for too long. There will have to be changes.[13]

The conversation went into neutral gear while the servant padded across the carpets and skilfully poured out our fifth cup of scented Yemeni coffee, and Yamani sipped morosely, listening to the voice of Diahann Carroll oozing smoothly out of the speaker above him.

It is always hard to be a moderate, [he said after a time] but it is particularly hard with American oil people. They are apt to take moderation for weakness. They know that we in Saudi Arabia and Kuwait can't be, don't want to be as violent as the Libyans. Gaddafi just draws his oil royalties and puts it in the bank, or uses it to suborn people. We spend it on public works, on education, on people. I think ours is the better way. I think those leaders with the large bank accounts, hoarding their money, are becoming not only too rich but too arrogant. It is having a bad effect on them. They think highly of the way they can use all their hoarded gold to destroy, and we who are moderates suffer from their arrogance. Especially if those towards whom we are being moderate misunderstand, and refuse to cooperate.[14]

He was asked what would happen if, at the forthcoming conference, the majors still continued to follow Aramco's lead and turned down all approaches for a participation agreement. He shook his coffee cup to indicate to the servant that he wanted no more, but there was something about the convulsiveness of the gesture that made one think that, for a moment, the cup he was shaking was an obstinate American oilman.

I don't expect to finish the negotiations at one session at Geneva [he said] but if the Anglo-Americans come to the conference table and they are still as stubborn-minded as they have been until now —if they give us a flat no to our participation proposals—then we will meet, OPEC, I mean, and we will decide what to do. And what I think we will do is force the companies to give us shares—force them by legislation. It would be unfortunate. It would cause trouble —and probably more trouble for them than for us. But we cannot afford to play into the hands of the militants, and we will have to show that we mean business.[15]

He was asked whether the major companies realized how close to a crunch situation they had reached. He shook his head.

In July 1967 [he said] I had a meeting with the four major companies which own Aramco in a small town in New York State. It was a time when they had reduced the posted price without warning, and completely unbalanced the Saudi budget. It was desperate. I went all the way to New York to meet them. I advised them to do something to alleviate our financial situation. I told them: 'If you don't do anything for us, you will have to pay for it in the end. So do something quickly. I am warning you.' Well, perhaps they laughed at me. Anyway, they didn't do anything, and the result was that they got OPEC, and they have been paying the price ever since, at least double the price, I think, they would have had to pay if they had been more amenable.[16]

He rose to his feet and paced for a moment across the carpets. There was a faint sound from beyond the filigreed division, and once more the elusive scent of French perfume. Finally, Yamani motioned me back to my seat, and squatting beside me, he said:

I think you will be seeing the Aramco people in Dhahran in a day or two. You can convey a message to them for me. It is this: 'I am warning you again. Do something about participation. If you do not, the price you will have to pay will be heavier than ever before.'[17]

He rose again to his feet, this time to accompany me to the door. It was dark over Riyadh now, and except for the glow of the floodlit water-tower there was no light to be seen. It was very silent.

'I hope they receive the message properly,' said Ahmad Zaki Yamani. 'But I am afraid sometimes, when oilmen think only from a money angle, they get blind. *Salaam Aleikum.*'

On 7 January 1972, I flew from Jiddah across the dun-coloured wastes of the Empty Quarter (it was cloudy and stormy most of the way) to Dhahran, on the Persian Gulf. For the last half-hour of the journey, from beyond Al Hasa Oasis all the way to the coast, the dunes and crags of the desert were crisscrossed with pipelines and smudged with the smoke and flame of gas flares. In 1971 alone those pipelines had carried 1,641,615,000 barrels of Aramco oil to the tank-farms around Ras Tanura, the loading port of the Gulf. When you added the oil that was pumped by the Japanese-owned Arabian Oil Company (65,348,000 barrels) and by the wells of the Getty Oil Company (33,762,000 barrels), it brought Saudi Arabia's output for 1971 to 1,741,000,000 barrels, or 4,769,863 barrels a day, making Saudi Arabia the greatest producer of oil in the whole Middle East, ahead of its two nearest rivals, Iran and Kuwait.

And out in the Gulf, like great whales breasting the cresting waves, the tankers queued up to go into the loading points, there to fill up with 300,000 tons of crude in 30 hours, and sail off to the markets, one every 45 minutes.

Dhahran looks like a small residential town in a Southern California valley, full of neat, air-conditioned bungalows with green lawns running down to the pavements, street lamps out of a Pasadena avenue, and oleander hedges. American, British, Saudi and Arabian officials and their families live side by side, and join each other at evening parties. There are schools, hospitals, clubs, playing fields and swimming-pools, and a great commissariat where the choices run from hamburgers to curry and marzipan cakes to pecan pie and ice cream. The schools, the education and the medical attention are free, and any Aramco employee can get an interest-free loan to build a house – in Dhahran, Al Khobar or nearby Dammam – after he has been with the organization a year.

A Saudi Arabian official of Aramco (trained at Syracuse, passionately anti-Nixon but fervently pro-company) showed me around the College of Petroleum which, with generous help from Aramco – to the tune of $14,515,000 – has become the best training school for oil technology in the Middle East. It lies in the shadow of the old derrick of Well Number 7, on a hill now known officially as Jabal Steineke, after the US geologist who first drilled oil there and founded Aramco's fortunes. We watched the budding Arab oil technicians and the statistics about Aramco began to roll off the Saudi official's lips:

Of the company's 10,353 employees, 8,630 are Saudi Arabians (883 are Americans);

Nearly 7,000 of the Saudis have been working for Aramco for more than 15 years;

Average salary of the Saudis in the company is $3,900 a year;*

Seven thousand five hundred and seventy-eight Saudis have built their houses through Aramco's home-loan programme;

Two hundred Saudis are sent every two years (at company expense) to the US on post-graduate courses;

The 44 schools the Saudis' children attend have cost the company $35,400,000.

We pump more oil than any other company in the Middle East [said the official]. We market it more efficiently than any other company. We pay more to our government than any company in any other state.† And we are Saudi Arabian, registered and incorporated inside the Saudi Arabian state.

How do you tell a company which has 'Saudized' itself so thoroughly, and runs so efficiently, that it must henceforth cede twenty per cent of its share – and eventually more – to the state? That it must hand over a fifth of its control not only of its action inside the Saudi state, but all the way downstream to the gas stations on the freeways of the world?

Ahmad Zaki Yamani can call it anything he likes, participation or whatever [said an American official of the company], but it still sounds like nationalization to us. Twenty per cent nationalization to begin with, and then a gradual step-up of the takeover, year by year, until they have control of the whole operation. It just isn't on. Even if Aramco accepted the proposition, can you see our parent companies doing so? Can you see Jersey, Socal, Texaco or Mobil—especially Mobil—letting them in on their operations?[18]

I explain something which he no doubt already knows: that what Yamani and other participation advocates envisage is an

* Ahmad Zaki Yamani told the author of an official survey made by an independent investigation team of the priorities of Aramco's Saudi employees. 'Number One: They wanted the company to give them more education. Number Two: They wanted the company to give their children a proper education. Number Three: They want raises in salaries. They are the only workers in the Middle East oil industry who put education before more money.'

† $1,866 million in 1971.

327

arrangement rather more complicated but somewhat less drastic than he was claiming. The Saudi minister put it in these words:

You see, we do not want to get the oil merely for its own sake. We do not want just to obtain, say, 20 per cent of Aramco's production, then 30 per cent another year, and so on. What would we do with all that oil? If we went and dumped it on the market, we would be committing suicide. So we have to have an arrangement with the Aramco owners, jointly or severally, whereby:
1 Part of our entitlement of oil production would be marketed jointly with them through joint ventures downstream; and this is a business matter.
2 Another part would be marketed by them on our behalf on a commission basis.
3 A third part would be sold by us acting alone through the various ways and means at our disposal.
So I think all these elements should go together if we are to have a healthy, stable situation.[19]

The American shook his head. 'No matter how you slice it,' he said, 'it's still nationalization. And that's something we don't believe in. We're Americans, remember? Free enterprise, and all that. We'll never accept a government, not even the American government, as our partner.'

In that case, I suggested, Aramco and all the other majors were in for a bumpy ride when the Geneva negotiations on participation started in fifteen days' time, for the oil ministers – with Yamani as their spokesman – were determined not to take no as an answer.

'We will legislate,' Yamani had said.

The night before I flew out of Arabia, a senior Saudi official of Aramco asked me to a party at his house in Dhahran. Faysal al Bassam spent his early childhood wandering the fringes of the Empty Quarter with his father's bedouin tribe, and is still so steeped in desert lore that those of his American fellow workers whose favourite weekend pastime is picknicking and exploring the secrets of the vast Arabian sand-sea consider that their trip is made when he consents to come along. He holds degrees in engineering and geology from US universities, speaks fluent English and French, and is regarded as one of the most promising members of the company's executive staff. That, plus an easy charm and dark good looks, makes him the most eligible bachelor with Aramco, and there is much nail-nibbling in the harems as they worry whether a Saudi girl or an American will eventually get him.

Al Bassam had invited a good cross-section of Aramco's senior staff to meet us: a young American vice-president of Tapline (which carries Aramco's crude across the Arabian desert to the Mediterranean), an exuberant Bahraini who works for Bapco, the subsidiary of two of Aramco's parent companies, in the refineries on Bahrain Island, a Saudi geologist and his English wife, and a sprinkling of American, Lebanese and European secretaries. In the middle of an animated conversation and platefuls of Boeuf Stroganoff, the president of Aramco, Frank Jungers, arrived with his wife. What had started as a social evening was almost immediately transformed into something different, because Jungers, alerted by his officials, wanted to hear as full a version as possible of what Ahmad Zaki Yamani had told me was OPEC's attitude to participation.

What seemed surprising at the time was the poor liaison which evidently existed between Aramco and the Saudi government to the extent that the gravity of the participation question had to be conveyed to the Americans through the medium of a conversation with a visiting writer. Jungers seemed genuinely surprised when I mentioned the determination of Yamani and his OPEC colleagues to push the participation question to the limit, and that, in the event of a no from the major oil companies, 'We will legislate'.

But he then hastened to assure me that Aramco was by no means as intransigent as some of his junior officials had suggested. He was flying to Geneva for the meeting with OPEC representatives the following day.

'I can tell you that we have no intention of turning them down,' he said. 'You will find that we will be going a long way towards meeting them.'

But not, as it turned out, at the place where the oil states wanted to go.

The great weakness of the Middle East oil states has always been their inability to stand firm and together in an emergency, and their susceptibility to bribery and corruption. Even the formation of OPEC has not eliminated the tendency of member states to revert to wily oriental habits when the opportunity for a quick profit offers itself. Such was the case, for instance, after the Six-Day War in 1967, when the Arab oil states voted to boycott certain markets, and their fellow member, Iran, immediately stepped up production to meet the shortage, to its considerable profit. And all through the history of Middle East oil, one state or another has been played off against its rivals through the skilful manipulations of the major oil companies.

But in 1972 throughout the participation negotiations the oil states stood firm in the face of considerable temptation. Ironically enough, considering the noise and posturing of the militant Arab states whenever oil talks are in progress, it was the attitude of two of the most autocratic of the old regimes which kept the oil states united. And it was King Faisal of Saudi Arabia (with the cooperation of the Shah of Iran) who pushed the negotiations through to a successful conclusion.

It so happened that the expected confrontation failed to take place at Geneva in January 1972. At the last moment, Ahmad Zaki Yamani skilfully postponed it. The delegates from the OPEC states and the representatives of the major oil companies had gathered in Geneva for two main purposes: first, to decide at what rate the oil states should be compensated for the revenue they had lost through the devaluation of the US dollar in 1971, and second, to make a decision about participation. The dollar question was quickly and efficiently dealt with; the majors, having gone to Geneva expecting to pay more, agreed to do so.

But the resigned amiability which the majors displayed over the money question (they would, after all, simply pass the increases over to the consumers) changed to outright hostility when the participation question reached the agenda. The Americans, particularly the four parent companies owning Aramco, made it quite clear to their more dovelike British, Dutch and French colleagues that they were determined to take a hard line. It was a situation made for the troublemakers among OPEC's more militant members, who could easily exploit American intransigence and force a public showdown from which, for political reasons, it might be impossible for the oil states to retreat. With the help of OPEC's secretary, a supple-minded Iraqi named Nadim Pachachi, and the lobbying of the Iranian delegate, Mansoor Froozan, Yamani manoeuvred the member states into agreeing to let him mastermind the negotiations on his own. At the same time the British, Dutch and French agreed to allow the four most stubborn US companies, Standard of New Jersey, Standard of California, Texaco and Mobil, to represent them in the talks with Yamani. Since they are the parent companies of Aramco, it had become a straight Saudi-Aramco confrontation. Yamani set the date for the talks between the two as 1 February 1972. He set the place as Jiddah, in Saudi Arabia, where a close government control was kept over the press and nothing of the negotiations would leak out of the conference room except what he wanted to appear.

It was at this stage that the US companies, using Frank Jungers of Aramco as their front man, decided to try a last trick to break the solid OPEC line-up on participation.

Jiddah is the most raffish of Saudi Arabia's cities. Its old wooden houses, with their jutting latticed balconies, are being jostled now by office buildings which squat moistly beside them on the shores of the Red Sea. February is the month when the pilgrims to the great haj are streaming back into the port from Mecca, forty miles up the road, and the bazaars along the waterfront are thick with Muslims from the far corners of the earth, buying up souvenirs before taking their chartered planes and ships back home. One of the small pleasures of attending the Aramco-Saudi confrontation was to watch the preoccupied businesslike expressions on the faces of the American oilmen soften ever so slightly as they pushed their way to the morning conferences through the merry hordes of giggling, frizz-haired girls from Nigeria, nomads from the Sudan, bare-dugged old ladies from Abidjan, and smiling sari-ed men and women from Bali, the Philippines, Malaysia and India. Here were human beings at peace with themselves, happy and relaxed, and even momentary contact with them was a relief from the tense business in hand.

For the Americans had decided on a bluff, and they were nervous about whether they would get away with it. How best could they tempt Saudi Arabia to break the solid ranks of OPEC ranged against them? The American delegates had pondered, planned and argued long into the warm nights in search of an answer, and they had come up with a proposition which they believed would get them off the hook.

At their meeting with Ahmad Zaki Yamani, on 15 February, they offered him not twenty per cent but fifty per cent in a participation project with them. The only snag was that the offer was not for Aramco's functioning operations, but for developing and operating certain areas of Saudi Arabia where the presence of oil had been proved but not yet developed. Even so, for the Saudis it must have been a tempting offer, for Aramco was sitting on vast reservoirs of untapped oil which only that enormously rich organization had the capital to develop. And the US delegates made it plain to Yamani that if he accepted the offer on behalf of Saudi Arabia, they would make similar approaches to Iran, where the Anglo-American consortium also controlled great stretches of proved but un-exploited reserves. As one authoritative source in the Arab world put it:

On the face of it such an offer might have had a certain attrac-tion for Saudi Arabia and Iran where plenty of such undeveloped structures exist in the companies' acreage.[20]

And had not Yamani himself approached Aramco on two separate occasions with proposals for fifty-fifty participation deals in Saudi Arabia? Indeed he had. But on both those occasions he was not representing the whole of the OPEC states, nor was there any danger, if they had gone through, that participation deals in other oil states would be compromised. With the exception of Iran, none of the other states had viable areas which could be exploited under a fifty-fifty deal. Their hope for participation lay in the OPEC demand for a twenty per cent share of existing operations, and acceptance of the Aramco offer would sabotage their chances.

So Yamani rejected the American offer.

'It may be reckoned a tribute to OPEC solidarity that even the potential beneficiaries rejected the proposal out of hand,' the Middle East Economic Survey reported.[21]

It is the moment when a boxer throws his master punch and misses that he is at his most vulnerable, and so it was for the Americans. On the evening of 15 February, a few hours after he had rejected Aramco's offer out of hand, Ahmad Zaki Yamani drove out of the old town of Jiddah to the modest palace King Faisal has built for himself along the flat, humid shores of the Red Sea. Contact had already been made with Teheran and the shah had let it be known that he approved of the Saudi rejection. Now Yamani decided that the moment had arrived for him to come in with a master punch of his own. He talked for a long time with the Saudi monarch, and he came away well satisfied with the results of his audience. Once before (as Howard Page of Standard of New Jersey had cause to remember) Yamani had used King Faisal to back up his tactics. Now he was about the play the king again.

That night the US delegates were informed that they would be expected to attend an extraordinary session the following morning, and when they were assembled, the Saudi oil minister wasted few words. He reminded them that the purpose of the conference in Jiddah was to find a solution to the crisis between the oil states and the oil companies over participation, and that no progress had been made. But the deadlock must be broken, and to emphasize the urgency of the situation, Yamani proposed to read a message to the delegates. It came, he said, from His Majesty the King, and this is what it said:

Gentlemen, the implementation of effective participation is imperative, and we expect the companies to cooperate with us with a view to reaching a satisfactory settlement. They should not oblige us to take measures in order to put into effect the implementation of participation.[22]

It was another way of saying what Yamani had said six weeks earlier: 'Give in, or we will legislate.' Only this time the ultimatum was given by the king himself, and his prestige, as well as that of his government, was now involved.

All the same, it took another three weeks for the Americans to come round, and Frank Jungers had to use his most persuasive arguments to convince some of the more recalcitrant among his parent companies that this was the end of the road. But two weeks later, after a succession of emergency conferences in Beirut, New York, San Francisco and Houston, it was done. On 11 March 1972, Ahmad Zaki Yamani announced:

Last night [i.e. 10 March] I received a letter from Aramco indicating the acceptance by Aramco and its shareholders of the principle of participation to the extent of a minimum of 20 per cent. The details of this question will be subject to further negotiations which will start in a week's time in Riyadh, Saudi Arabia. The OPEC extraordinary conference will be informed today of this step taken by Aramco and its shareholders.[23]

It was done. Aramco had accepted and the rest would follow. Admittedly it was only the principle of participation which had been accepted so far, but everyone knew there could be no turning back. One oil expert in the Middle East commented:

One should not, of course, underestimate the formidable problems that still remain to be settled—notably the price to be paid for the government's 20 per cent share . . . Nevertheless, basically the Rubicon has been crossed, and it can safely be affirmed that the final movement towards host-government participation in the world's largest concentration of low-cost oil reserves has been well and truly launched.[24]

It was, as they remarked in the Middle East, 'a well-deserved personal triumph for Sheikh Ahmad Zaki Yamani, who for many years now has been the foremost champion of the idea of participation both in Saudi Arabia and in the general framework of OPEC.'[25]

Things would never be the same in the Middle East oil business, and in the eyes of some veteran British and American oilmen it was the beginning of the end. Certainly, it would be different. But with a dynamic character like Yamani henceforward sitting on that Aramco executive committee in New York ('I want to be there,' he had once said), it wouldn't all be bad.

28 · The Bottom of the Barrel?

So we reach the present day in this story of Middle East oil. What happens now? It is not likely that the successors of Frank Holmes, Knox D'Arcy, Calouste Gulbenkian, Jean Paul Getty, Max Steineke and Enrico Mattei will be as flamboyant as those characters who have figured in this story, but the events in which they will undoubtedly find themselves involved will almost certainly be as spectacular, hazardous and unexpected.

The Middle East oil industry is changing fast, and no matter what else happens, the changes are not going to make life too comfortable for the Western oilmen who run the concessions or have invested scores of millions of pounds in their future. Nor are their parent countries likely to remain aloof from the problems which are now looming on the Middle East horizon. Once upon a time the US was interested in exploiting the oil-wells of Arabia only for the money that could be made out of them. Nowadays the Americans have become much more interested in the oil itself, for their own use, and this cannot but cause the deepest concern to the Europeans and Japanese who have, until now, regarded the Middle East as the main source of their energy supplies.

'In 1970, the US consumed 710 million tons (30 per cent of world consumption) or 15 million barrels a day,' says a recent US report.

By 1980 it will demand between 20 and 25 million barrels a day, but US production, now around 10 million, will rise to only 11 million from presently known reserves. Alaska could provide 2,000,000 barrels a day if the currently planned (and hotly contested) pipeline is built; newly found offshore deposits could swell output if environmental objections could be met. Even so, industry sources predict that there will be a gap.[1]

Until new sources of energy have been developed, that gap can only be filled by buying more oil, principally from the Middle East. And the Middle East price is going up, both politically and monetarily. To fill America's needs, Ahmad Zaki Yamani of Saudi Arabia has proposed a preferential oil agreement between the US and his own country whereby priority would be given to America's fuel requirements over all other customers in return for an 'open door' for Saudi investment in the US. But while this might solve America's problem, it would create even graver ones for Europe and Japan. Today most oil from Arab and Iranian fields goes to Western Europe or Asia. Exports of crude oil from Saudi Arabia in 1971, for instance, were divided as follows: 48 per cent to Western Europe, 37 per cent to Asia, Africa and Australia, 10 per cent to North America and 5 per cent elsewhere. At a moment when several Arab states – Libya and Kuwait in particular – are deliberately cutting down production to hold on to diminishing reserves, oilmen fear the advent of the United States into the Middle East market as a consumer could only starve Europe or drive up the price that everyone will have to pay.

The French state oil group, Elf/ERAP, expressed the concern of its British and European collaborators in an editorial in its monthly bulletin in October 1972, shortly after Ahmad Zaki Yamani had proposed his 'marriage', as he called it, with the United States.

'If completed at a government level,' the editorial commented, 'the preferential oil agreement between the US and Saudi Arabia will be a primordial event for world energy supplies. The "catholic marriage" – to use an expression favoured by the Saudi minister – between the biggest consumer and the largest producer of energy in the world; the tapping by the US of the largest share of available reserves, at a time when sources of cheap oil are becoming rare; the strengthening of the security of the US's supply through Saudi capital invested in its territory; all this is clearly an excellent deal for the two partners, perhaps the greatest business marriage in history.'[2]

But what of Saudi Arabia's traditional customers from Europe and Japan?

'Those who will be invited to watch cannot but be worried,' the editorial continued. 'They know they will have to pay the cost: Europe supplies 58 per cent of the oil income of the Gulf states and Japan 16 per cent, as against only 12 per cent up till now from the future partner. It will be a strong inducement to raise the price of crude oil if the money paid by Europe and Japan to the [American] producing companies finds its way back to the country of origin of these companies as investment via the producing state, and reinforces their power there. Who could then maintain that the companies which handle Arabia's production are impartial intermediaries between that country and European consumers.'[3]

There is no real fear in Britain and Western Europe that oil supplies are running out, and most experts would disagree with David Freeman, a former energy expert for President Nixon, who said recently that 'our rates of consumption are so large that we can see the bottom of the barrel.'[4] Despite the fact that Kuwait and Libya may well have less reserves of oil than they believed a year or so ago, the reserves in other parts of the Middle East do not cease to grow. Iran alone, which produced 52 million tons of crude oil in 1960, 191.7 million tons in 1970, and 227 million tons in 1971, is planning to increase production to 300 million tons a year in the next decade. Saudi Arabia has even greater expansion plans. Ahmad Zaki Yamani announced in New York in October 1972 that his country plans to drive production up by 1980 to an astounding 20 million barrels a day, or 1,000 million tons a year.

The oil will be there, and no matter how high a price Europe may have to pay for it, there is no doubt that the money will be found. As long as there are factories to be fuelled, automobiles to be lubricated and houses to be heated, the price will be paid in spite of restrictive agreements and OPEC demands, for petroleum has become as much the drug of Western civilization as cigarettes and alcohol, and the addicts will go on paying. But there is also another price which may be demanded of them, the result of the increasing tendency of Arab militants to attempt to use oil as a political weapon.

'Oil will do what no Arab government has been able to do in imposing upon the West a realistic attitude on the Arab Middle

East issue,' wrote a Kuwaiti director of the Anglo-American Kuwait Oil Company in November 1972.⁵ He left his readers in no doubt that what he meant was that the West's thirst for energy for its people and machines would force the adoption of a pro-Arab stance in its attitude towards the conflict with Israel. And who could say he was wrong?

It was not simply antipathy towards the Israelis that persuaded General de Gaulle to swing France's policy behind the Arabs in the last years of his regime, and has impelled his successor, President Georges Pompidou, to curry Arab favours. Both leaders were prepared to toe the Arab line in the hope that one day it would bring dividends in the form of oil concessions, and in the summer of 1972 this policy paid off. When Iraq nationalized IPC's holdings in June 1972 the confiscation was directed at the American, British and Dutch partners of the combine. The French partner, Compagnie Française des Petroles, was singled out for special exemption, and invited to discuss ways and means of taking over from her erstwhile con-frères. France was being rewarded and given a 'privileged position', said Saddim Hussein, vice-president of the Iraqi Revolutionary Committee, because of the 'just policy that France follows concerning Arab causes and specifically the Palestinian cause against Israel.' He added that any country which adopted a similar attitude 'will be able to profit from a privileged situation, not only in Iraq but also in all the Arab countries.'⁶

His words must have stirred groans of dismay from Western oilmen in the Middle East, and as I read them there re-echoed through my head what many an executive, British or American, had said whenever we spoke politics during my stay in the oil states:

Why does the United States have to keep backing Israel, anyway? What does it get them—except the Jewish vote in New York in an election year? Why don't Washington and London and the Hague behave like France, and subordinate Middle East policy to self-interest? A pro-Arab attitude in the State Department or Whitehall could get us all everywhere.

There is no doubt at all that it could buy time for the major oil companies, and stave off the day when complete control of their fields inside the producing countries is taken over by national (Kuwaiti, Saudi Arabian, Abu Dhabian, Iraqi) oil companies. But oil executives in Britain and America, Foreign Office strategists, State Department planners, MPs and Congressmen are almost certainly fooling themselves if they believe that a blindly pro-Arab (and its concomitant, an anti-Israel) policy will save them for anything more than a temporary period.

Iran, which demands no anti-Israel or pro-Arab line from its foreign concessionaires,* is nonetheless determined to take over the Anglo-American consortium and run the whole enterprise herself within a foreseeable period of years, for no other reason than that it is now run by foreigners and the Iranians wish to run it themselves.

In March 1972, the Shah of Iran gave an interview to a French oil journal during the course of which he was asked:

'Do you intend eventually to take your oil affairs into your own hands and sell your oil products directly?'

He replied:

There is no doubt that is our final goal. We are currently studying this question, notably with our neighbours and friends in Kuwait and Saudi Arabia. As far as we Iranians are concerned, we are too sophisticated to act purely and simply as nationalists. Twenty years ago we suffered a very serious reverse. It was a hard and lasting lesson. We still have abundant oil for a good fifty years, while other countries see their reserves drying up. We are also trying to set up long-term plans. Already we are thinking of 1979, the date of the end of the first period of the [Anglo-American] oil agreement. Obviously we could accept another three further five year periods [as set out in the agreement] but that would be the easy way out.[7]

The shah has made it clear that though he approves Ahmad Zaki Yamani's campaign for participation in the operations of the major oil companies, such an arrangement is of no direct concern of Iran, which secured a participation agreement (albeit an unsatisfactory one) from the consortium under the agreement of 1954. He regards participation as no more than a temporary expedient to make use of the foreign companies

---

* Much of its crude oil from the National Iranian Oil Company reserves is, in fact, being shipped to Elath, the South Israel port, and transported by the new 47 inch pipeline to the Mediterranean, whence it is taken by tankers to Eastern Europe.

and their facilities while building up a technical and sales staff and marketing expertise.

'I am thinking of the 40 million Iranians who will be alive in the year 2000,' the shah said; 'that is why I say that the ideal final solution is for us to take our oil into our own hands, by-passing the intermediaries (the majors) to a greater and greater degree. Today this might come as a shock, but I do not think that when it comes the producers and hundreds of millions of consumers will complain about it.'[8] *

If the ruler of a land with close political as well as economic ties with the US and Western Europe is nevertheless preparing to take over the role of the major companies in his oilfields, is it likely that similar plans can be staved off in the other Middle East states, even by the most servile policy towards the Arabs?

Some oilmen would probably reply to that by pointing out that the Anglo-American consortium's concession in Iran expires in 1979, whereas those in Kuwait, Saudi Arabia and the Gulf Emirates run to the end of the century, and some of them well beyond that. To Western oilmen to whom a contract is sacred that may be a comforting fact. But there are few Arab governments today which consider themselves bound by the concession agreements they signed with the major companies before World War II, and if they stick to them for the moment it is only because the time is not ripe, the technicians are not numerous enough, and the dispositions have not been made, for confiscation. Nadim Pachachi, Secretary General of OPEC, expressed the views of most of his members when he said:

As you are aware, the basis of a valid contract is the free will of the two parties. Most of the oil concessions granted in the Middle East before and after World War Two were concluded by states under the mandate or influence of a colonial power. When the IPC agreement was concluded in 1925, Iraq was under British mandate. It is a well-known fact that the frontier commission appointed by the League of Nations to settle the dispute between Iraq and Turkey over the Vilayet of Mosul refused to give a decision in favour of Iraq before an oil concession was concluded with IPC. The British government would not allow the ratification of the new constitution before the oil concession was granted.[9]

* The shah confirmed Iran's intention of taking over the consortium's concession when it expires in 1979, in a statement issued on 23 January 1973.

There is therefore nothing, Pachachi maintains, to prevent a state from repudiating a concession or altering it whenever there is 'a strong and valid reason for the revision.' But what constitutes a 'strong and valid reason'? Mouammar Gaddafi nationalized British Petroleum's concession in Libya because Britain failed to prevent Iran from seizing offshore islands in the Persian Gulf, thousands of miles away. Iraq nationalized IPC's northern holdings because, it claimed, the company was holding up production.

Yet, curiously enough, not a single Middle East oil state has yet nationalized any US oil company, despite the fact that American policy in support of Israel continues to bedevil relations between the Arab states and Washington. If British inaction in the Gulf is sufficient reason for seizing BP's assets in Libya, why hasn't Gaddafi seized Esso's even bigger concession in retaliation for the US's solid support of Israel?

Could it be because Gaddafi believed he had enough technicians to handle BP's operations, but lacks the experts to handle Esso as well? Or simply because he can gain easy propaganda returns from twisting the tail of a comparatively helpless British lion, but finds the US too big an opponent and too useful as a market and as a provider of technical know-how to wish to take that step?

There are students of the Middle East oil situation who believe that it would be better for everyone, particularly the consumers, if the OPEC oil states did seize and nationalize all oil operations inside their countries, and sooner rather than later. James Akins, Director of the Office of Fuels and Energy in the US State Department, discussed the implications of such a move in a speech he made on 2 June 1972, at a meeting of the Eighth Arab Petroleum Congress in Algiers. He made the point that nationalization might not necessarily be a disaster for the parent countries of the Western oil companies. He pointed out that if all US investments in oil were taken over, including Canada, Venezuela, non-OPEC Africa as well as the Middle East, it would mean a loss of $700 million to America's balance of payments. British and Dutch losses would be about two thirds of that.

But what would the Arabs gain?

'Arab income from oil in 1970 was $4,800 million,' he said.

Based on increases in taxes already agreed upon by OPEC and the companies, and on published figures for increases in production, this will rise to over $12,000 million by 1975. And by 1980, assuming further price increases, the annual income has been estimated at somewhere between $30,000 million and $50,000 million. I should venture that if production of all the major (US, British and Japanese) companies in the Arab world were nationalized, the additional benefit to the Arabs would be well under $1,000 million a year. And if the consumers started crash programmes to develop other energy forms, it would be substantially less.[10]

And what would the consumer lose? Akins points out that he might even gain. The argument of those who believe that nationalization should be encouraged 'is that once the oil companies' buffer is removed, OPEC's solid front will disappear and the national companies will compete among each other for a greater share of the markets, and the prices will go down to one dollar a barrel in the Gulf.'[11]

To the major companies, of course, Akins' remarks bring no comfort, and they view the prospect of the nationalization of their enterprises not only as an unmitigated disaster but as a fate they do not deserve. They cite the enormous capital investment they have made (and are planning to make) in the Middle East, and insist that no compensation offered could possibly match the effort and expenditure they have put into their concessions.

Indeed, their efforts have been vast and unremitting, as this book has tried to show. But it has also been rewarding. The majors have recovered their investments many times over, and they go on making profits on a munificent scale. The US Department of Commerce, for instance, reporting on direct US investments in 1970, pointed out that the net assets of the petroleum industry in the Middle East amounted to $1,466 million. The yield from those assets in 1970 alone was $1,161 million, representing an annual return of 79.2 per cent. On the other hand, US investments in overseas mining and smelting industries yielded 13.5 per cent, and in the same year (1970) the yield from overseas manufacturing companies was 10.2 per cent. British figures are not likely to be much different.

Those fatalists in the oil industry who have begun to accept nationalization of oil concessions in the Middle East believe it could get the US off the hook, politically speaking, and relieve pressures on Europe and Japan. It would depoliticize petroleum and restore it as an article of merchandise to be sold at the going price. It would take the oilfields out of the Arab-Israeli conflict. Moreover, any militant who thereupon suggested that the oil states, controlling their own supplies through their own companies, should immediately begin a boycott of petroleum to those Western nations pursuing 'anti-Arab' policies would crash with a considerable impact against the facts of commercial life. In business on their own, unbolstered by the major oil companies' posted price system, no oil state would dare to initiate an oil boycott against any consumer nation with the money to buy. In all the Middle East there are not enough storage tanks to hold the oil that would immediately become surplus if such a boycott were attempted. The deserts would start to run with crude. The gas stoves of Kuwait and Eastern Arabia would splutter and go out. The desalination plants would cease to function, the air conditioning plants would stop, and the water tanks would run dry. And those militants advocating a massive cutback in production in order to create a 'sellers' market would then be faced by the harsh facts about oilfields: you can diminish the flow from the wells or shut them down completely, but if you want to use them again you should close them down only for a limited time. A well closed down too long may not only be impossible to reactivate, but it may also play havoc with the rest of the field.

The Arab states might well come to the conclusion that even the pleasure of seeing Western Europe and the United States squirm under a boycott is not worth the destruction of a multi-million dollar oilfield.

Nationalization of the major oil companies' concessions in the OPEC states of the Middle East would, it is true, effectively destroy the international oil cartel. It would also bring to an end the artificial price-structure for petroleum which, between them, the OPEC countries and the cartel have created. It is not likely that many consumers would grieve over that. The companies would become 'off-takers' searching the world for the cheapest oil they can buy for their markets. The OPEC nations would lose the buffer organization which sells their products for them at an inflated price. And only the consumers would gain.

It is for this reason, of course, that, despite Iraq, despite

343

Gaddafi, despite the militants' cry of 'Arab oil for the Arabs' and 'Hands off Iranian oil', the international oil cartel is not likely to be put to death just yet by the OPEC states. So long as it has a role to fulfil on behalf of the producing countries, so long will the cartel be allowed to go on operating, and so long will the major companies continue to make their heady profits.

In forcing them to accept participation, Ahmad Zaki Yamani has saved them from extinction. One wonders why they ever objected to it, for, as Maurice Chevalier said about those who grumble about old age, consider the nature of the alternative. True, in the days to come, the great British and American combines will find themselves giving larger and larger shares to their host governments, and paying bigger and bigger royalties and higher taxes. But if their past performance is anything to go by, they will simply pass on the increases to the consumer.

Perhaps at the end of the present decade, or maybe at the end of the century, power will have passed out of their hands into that of the producing countries. Gulf, Esso, BP, Shell and Mobil will still be names in neon lights on the petrol stations along the world's highways,* but their names will have disappeared from the wellheads in favour of NIOC, INOC, Petromin, KNOC, and ADNOC.†

The directors of the ousted companies will bewail the lack of gratitude of the governments in whose deserts they sweated and pioneered, and their shareholders will no doubt gnash their teeth over the smallness of the compensation.

But those who have thought about this story while reading these pages may come to the conclusion that, all things considered, pioneers, companies and shareholders have all had a pretty good run for our money.

* Not Esso, perhaps. Its parent company, Standard of New Jersey, changed its name to Exxon Corporation in the autumn of 1972.

† National Iranian Oil Company, Iraq National Oil Company, Saudi Arabian Oil Company, Kuwait National Oil Company, Abu Dhabi National Oil Company.

Maps

Maps

Map 1 – Oil wells of the Persian Gulf

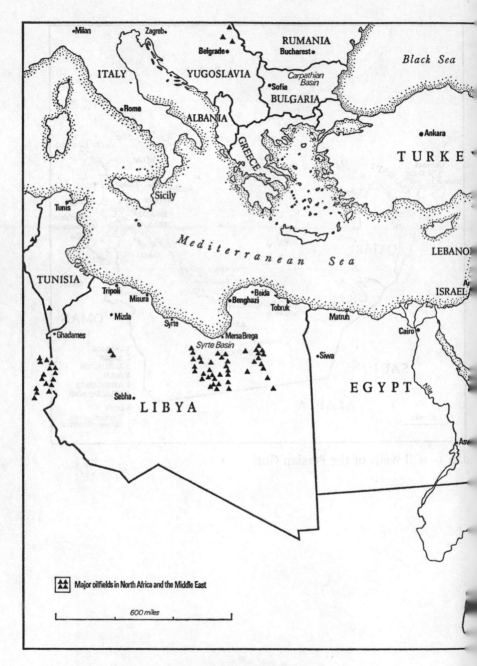

Map 2 – North Africa and the Middle East

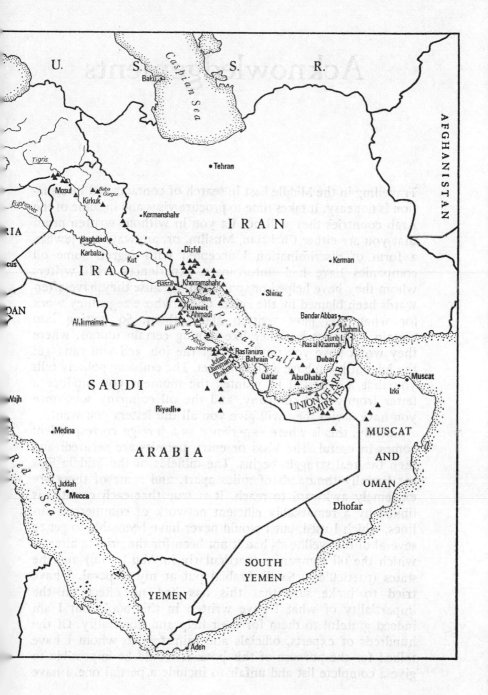

Acknowledgements

Travelling in the Middle East in search of contacts and information is not easy. It takes time to procure visas and in some of the Arab countries they will not let you in without written proof that you are either Christian, Muslim, or, anyway, not Jewish, a form of discrimination I accepted with regret. Some oil companies have had unfortunate experiences with writers whom they have helped or sponsored, because they have afterwards been blamed by the rulers under whose aegis they work for what was later written about them. So getting into certain countries is rather like joining certain unions, where they won't have you until you get the job, and you can't get the job until you have a union ticket. The embassy politely tells you that a visa will be granted the moment you produce a letter from an oil company, and the oil company says once you have a visa they will give you all the letters you want.

Luckily, this is where experience as a foreign correspondent comes in useful. The visas or entry permits are secured: and then the real struggle begins. The oilfields of the Middle East are literally thousands of miles apart, and some of them are extremely awkward to reach. It is true that each country is linked by a remarkably efficient network of commercial airlines, which I used, but I would never have been able to get to several of the wellheads had it not been for the private aircraft which the oil companies (particularly in Abu Dhabi) and the states (particularly Saudi Arabia) put at my disposal. I have tried to make sure that this has had no effect on the impartiality of what I have written in this book, but I am indeed grateful to them for their help and hospitality. Of the hundreds of experts, officials and ministers to whom I have talked for the purpose of this book it would be impossible to give a complete list and unfair to include a partial one. I have therefore decided to give thanks by name only to those whose aid both before and during my travels went far beyond the call of duty.

I must, for instance, particularly single out the continuous advice and information which was put at my disposal by Mr Fuad W. Itayim. Mr Itayim is editor and publisher in Beirut of the knowledgeable, authoritative and influential bulletins of the *Middle East Economic Survey*, a weekly review of news and views on Middle East oil. He placed his office and his files at my service, and he and his staff were always ready to answer my questions. Furthermore, the news editor of *MEES*, Mr I. N. N. Seymour, whom I encountered in several remote parts of the Persian Gulf, was always ready with news and with introductions to oil sheikhs and oilmen, with scores of whom he is on special terms of trust and friendship. I am most grateful both to him and to Mr Itayim.

In the United States, in both his office and his home, Mr Howard Page, that wise spokesman of the oil industry, took hours away from his other occupations to retrace the critical moments in the story of Middle East oil. In Saudi Arabia, Sheikh Ahmad Zaki Yamani, the oil minister, did likewise, at moments when every ruler in the Middle East was seeking his counsel. And it was during the same period of great crisis, when the whole future of Western oil interests in the Middle East was in the balance, that Mr Frank Jungers, president of Aramco, talked at length to me about his problems.

A few other names whose services, for space reasons, I will not specify:

Mr Shafiq Ombargi, Mr Farid Riszk, Mr Ihsen Hijazi, Mr Sinclair Road, Mr Sulaiman Olayan in Beirut; Mr Harry McDonald, Mr Faysal al Bassam, Mr Bill Mulligan, Mr Pete Speers in Dhahran; Mr Mohammad Almana in Al Khobar; Mr Abdullah Hashim in Dammam; Mr Ghaleb Abu al Faraj, Mr Abdullah al Ghanim and Mr Fahd al Sedayri in Riyadh; Dr Tom Temperley and Mr Abbas Sindi in Jiddah; Mr Tony Leigh-Morgan, Mr Peter Rawlston, and Mr Richard I. L. Stephens in Kuwait; Mr Vittorio Gismondi in Dubai; Mr Ahmed Obaidli, Mr Zaki Nusseibeh, Mr Ahmed Hijji, Mr R. A. de Freitas, Mr Geoffrey W. G. Ball, Dr Alan Horan, Mr Robert Bell, Mr John Millward in Abu Dhabi; Mr Ronald F. Rosner and Mr Joseph Johnston in New York; Mr Julius Edwardes and Mr Geoffrey

Keating in London (and, in the case of the latter, in several parts of the Middle East); Dr Wendell Phillips in London and New York; and Dr Nadim Pachachi and various members of OPEC and OAPEC in all parts of the Middle East. In addition, I should like to thank Mrs Joan St George Saunders for allowing me to study and quote from the unpublished history of British Petroleum written by her husband, the late Hilary St George Saunders.

Finally there was my travelling companion, who not only, as usual, looked, listened, and took notes, but also, in the blackest moments, made the desert smile. To her my special thanks and to the rest, my grateful salaams.

Source Notes

CHAPTER 2

1 Longhurst, Henry, *Adventure in Oil* (London 1959).
2 Zischka, Antoine, *La Guerre Secrète pour le Pétrole* (Paris 1933).
3 Quoted from Saunders, Hilary St. George, *History of the British Petroleum Company* (unpublished). This was lent to the author by his widow.

CHAPTER 3

1 Hewins, Ralph, *Mr Five Per Cent: The Biography of Calouste Gulbenkian* (London 1957).
2 Lord Curzon, *House of Lords debates*, 5th S.XL (1920).
3 Gulbenkian, Nubar, *Pantaraxia* (London 1965).
4 I recommend a study of Nubar Gulbenkian's autobiography, the biography of his father by Ralph Hewins, which Nubar authorized, and the autobiography of Sir Henri Deterding for the often unsavoury details.
5 Hewins, *Mr Five Per Cent*.

CHAPTER 4

1 Rihani, Ameen, *Ibn Sauod of Arabia: His People and His Land* (London 1928).
2 Rihani, *Ibn Sauod of Arabia*.
3 Rihani, *Ibn Sauod of Arabia*.
4 Rihani, *Ibn Sauod of Arabia*.
5 Rihani, *Ibn Sauod of Arabia*.
6 Rihani, *Ibn Sauod of Arabia*.
7 Rihani, *Ibn Sauod of Arabia*.

CHAPTER 5

1 Philby, H. St. John, *Arabian Jubilee* (London 1954).
2 Philby, *Arabian Jubilee*.
3 Stegner, Wallace, *Discovery!* (Beirut 1971).
4 Stegner, *Discovery!*

353

5 Longrigg, Stephen H., *Oil in the Middle East*, 2nd ed. (London 1968).
6 H. St. John Philby, *Arabian Oil Ventures* (Washington 1950).
7 H. St. John Philby, *Arabian Oil Ventures*.
8 Stegner, *Discovery!*

CHAPTER 6

1 Quoted by Hewins, Ralph, *The Golden Dream* (London 1963).
2 Violet Dickinson, *40 Years in Kuwait* (London 1971).
3 Quoted by Hewins, *The Golden Dream*.
4 Hearings on *American Petroleum Interests in Foreign Countries*, memorandum submitted by Charles Raynor, petroleum adviser to the State Department, 79th Congress, 1st session (27–28 June 1945).
5 Hearings on *American Petroleum Interests in Foreign Countries*.
6 Ward, T. E., *Negotiations for Oil Concessions in Bahrain, Al Hasa, The Neutral Zone, Qatar and Kuwait* (printed privately 1969). Quoted by Stocking, George W., *Middle East Oil* (Nashville, Tenn. 1970).
7 Longrigg, *Oil in the Middle East*.
8 Longrigg, *Oil in the Middle East*.

CHAPTER 7

1 Longrigg, *Oil in the Middle East*. Stephen Longrigg was a member of IPC's staff.
2 Quoted by A. Beaby-Thompson, *Oil Pioneer* (London 1961).
3 Beaby-Thompson, *Oil Pioneer*.

CHAPTER 8

1 From the notes Max Steineke was writing, and left unfinished, when he died in 1952.
2 From Max Steineke's notes.
3 I am grateful to Bill Mulligan and Pete Speers, in charge of the Aramco archives in Dhahran and to Wallace Stegner's official history *Discovery!* for helping me to detail this story.
4 From a document in the archives at Dhahran.

CHAPTER 9

1 Hearings on *American Petroleum Interests in Foreign Countries*, memorandum submitted by Charles Raynor.

CHAPTER 10

1 From Saunders Hilary St. George, *History of the British Petroleum Company* (unpublished).
2 In a conversation with the author in Ankara, Turkey, in 1943.

3 For what was happening in Baghdad, Freya Stark, *East is West*, (London 1945) is recommended. She was a refugee in the British Embassy and pays a warm tribute to the steps taken by the US ambassador, Paul Knabenshue, to protect British and British Indian citizens left behind, from the wrath of the Iraqi crowds.

4 Records of the Anglo-Iranian Company.

5 Records of the Anglo-Iranian Company.

6 Records of the Anglo-Iranian Company.

CHAPTER 11

1 Stegner, *Discovery!*

2 Stegner, *Discovery!*

3 Quoted during a Senate Inquiry, *Petroleum Arrangements with Saudi Arabia*, 80th Congress, (Washington DC 1948).

4 Senate Inquiry, *Petroleum Arrangements with Saudi Arabia*, Moffett quoted in memorandum submitted by W. S. S. Rodgers, Chairman, the Texas Company.

5 Senate Inquiry, *Petroleum Arrangements with Saudi Arabia*.

6 Senate Inquiry, *Petroleum Arrangements with Saudi Arabia*, p 24747.

7 Senate Inquiry, *Petroleum Arrangements with Saudi Arabia*, p 25415.

8 Senate Inquiry, *Petroleum Arrangements with Saudi Arabia*.

9 Testimony before the Senate Committee *To Investigate Oil Resources*, 79th Congress, 1st session, (Washington DC 1945).

10 Testimony before the Senate Committee *To Investigate Oil Resources*.

11 Senate Inquiry, *To Investigate Oil Resources*.

12 Senate Inquiry, *To Investigate Oil Resources*.

13 Senate Inquiry, *To Investigate Oil Resources*.

14 Senate Inquiry, *To Investigate Oil Resources*.

15 *Colliers Magazine* (2 December 1944).

CHAPTER 12

1 The proceedings of the Senate Small Business Committee (11 March 1946). Congressional Record (Senate), Part Two.

2 Eddy, W. A., *F. D. R. Meets Ibn Saud* (New York 1954).

3 Churchill Sir Winston S., *The Second World War* VI (London 1948–53).

4 Philby, *Arabian Jubilee*.

5 Larson, Henrietta M., Knowlton, Evelyn and Popple, Charles S., *New Horizons: History of the Standard Oil Co. (N.J.)* (New York 1971).

6 Gulbenkian, *Pantaraxia*.

7 Larson, Knowlton and Popple, *New Horizons*.

8 Larson, Knowlton and Popple, *New Horizons*.

9 Larson, Knowlton and Popple, *New Horizons*.

10 In a conversation with the author.

11 In a conversation with the author.

12 Larson, Knowlton and Popple, *New Horizons.*
13 Larson, Knowlton and Popple, *New Horizons.*
14 Larson, Knowlton and Popple, *New Horizons.*
15 Gulbenkian, *Pantaraxia.*
16 Gulbenkian, *Pantaraxia.*
17 Gulbenkian, *Pantaraxia.*
18 Gulbenkian, *Pantaraxia.*

CHAPTER 13

1 Getty, J. Paul, *My Life and Fortunes* (London 1963).
2 Report of De Golyer and MacNaughton Inc., Dallas, Texas,
   Petroleum Geologists and Consultants (20 March 1958).

CHAPTER 14

1 Van der Meulen, D., *The Wells of Ibn Saud* (London 1957).
2 Philby, *Arabian Jubilee.*
3 Van der Meulen, *The Wells of Ibn Saud.*
4 Howarth, David, *The Desert King* (London 1964).
5 Van der Meulen, *The Wells of Ibn Saud.*
6 Testimony before Senate Committee, *Joint Hearings on the
   Emergency Oil Lift and Related Oil Problems,* 85th Congress (1957).
7 Testimony before Senate Committee, *Joint Hearings on the
   Emergency Oil Lift.*
8 Quoted by F. A. Davies, Testimony before Senate Committee,
   Joint Hearings on the Emergency Oil Lift.
9 Testimony before Senate Committee, *Joint Hearings on the
   Emergency Oil Lift.*
10 Testimony before Senate Committee, *Joint Hearings on the
   Emergency Oil Lift.*
11 Howarth, *The Desert King.*
12 Testimony before Senate Committee, *Joint Hearings on the Emergency Oil Lift.*

CHAPTER 15

1 Quoted by Longrigg, *Oil in the Middle East.*
2 Quoted by Hewins, *Mr Five Per Cent.*
3 Longrigg, *Oil in the Middle East.*
4 Longhurst, *Adventure in Oil.*

CHAPTER 16

1 Arfa, General Hassan, *Under Five Shahs* (London 1964).
2 Arfa, *Under Five Shahs.*
3 Arfa, *Under Five Shahs.*
4 Longhurst, *Adventure in Oil.*
5 Statement recalled by friends of Dr Amini, now living 'in retirement' on his estates outside Teheran.

6 In a conversation with the author.
7 In a conversation with the author.
8 Frankel, P. H., *Mattei: Oil and Power Politics* (London 1966).
9 Frankel, *Mattei*.
10 In a conversation with the author. Mr Page does not appear to have been aware of Enrico Mattei's meeting with Lord Strathalmond of Anglo-Iranian in London during the Mossadeq crisis. It happened, of course, before the American companies were involved.
11 Frankel, *Mattei*.

CHAPTER 17

1 Boustead, Colonel Sir Hugh, *The Wind of Morning* (London 1971).
2 Boustead, *The Wind of Morning*.
3 Boustead, *The Wind of Morning*.
4 Account given by Mrs Beth Thoms, wife of Dr Wells Thoms, US missionary in Oman. Quoted by Phillips, Wendell, *Unknown Oman* (London 1966).
5 In a conversation with the author.
6 Phillips, *Unknown Oman*.
7 In a conversation with the author.
8 In a conversation with the author.
9 In a conversation with the author.

CHAPTER 18

1 Trevelyan, Sir Humphrey, *The Middle East in Revolution* (London 1970).
2 Quoted by Hewins, *A Golden Dream*.
3 In a conversation with the author.
4 The US oil industry laid on an emergency petroleum export programme to Europe to take care of the shortages. The resultant sharp rise in the price of domestic fuel in the US led to a Senate Inquiry which produced some interesting details of the interlocking nature of the international oil combines. See Testimony before Senate Committee, *Joint Hearings on the Emergency Oil Lift and Related Oil Problems*.
5 *New York Herald Tribune* (4 November 1957).
6 Stocking, *Middle East Oil*.
7 Article by Enrico Mattei in *Politique Étrangère*, Paris, 4 (1957).

CHAPTER 19

1 Photostat in the possession of Research and Translation Office, *Middle East Economic Survey*, Beirut, Lebanon. See also *Al-Sayyad*, Beirut (19 March 1959).
2 Longrigg, *Oil in the Middle East*.

3 Longrigg, *Oil in the Middle East.*
4 Stocking, *Middle East Oil.*
5 In a conversation with the author.
6 Trevelyan, *The Middle East in Revolution.*
7 Quoted in *Middle East Economic Survey*, Beirut (25 July 1958).
8 *Financial Times* (22 July 1958).
9 *The Times* (24 April 1959).
10 *Financial Times* (29–30 April 1959).
11 In a conversation with the author.
12 Stocking, *Middle East Oil.*
13 Quoted by Stocking, *Middle East Oil*, from minutes printed in *Platt's Oilgram News Service* (24 November 1961).
14 Interview with Muhammad Salman, Minister of Oil, *Le Commerce du Levant*, Beirut (2 February 1963).

CHAPTER 20

1 Quoted by Howard W. Page in a conversation with the author.
2 Quoted by Ahmad Yamani, at present Minister of Petroleum (1972) in Saudi Arabia, in a conversation with the author.
3 This account is based on the records of the Middle East Research and Publishing Center, dated 28 October, 4, 11 and 18 November 1960, and contemporary newspaper accounts.
4 Transcript from the Middle East Research and Publishing Center, Beirut.
5 Seymour, Ian, in *Middle East Economic Survey* (28 October 1960).
6 Seymour, in *Middle East Economic Survey.*
7 Quoted from interview in *Al Zaman*, Baghdad (17 April 1961).
8 Quoted by Stocking, *Middle East Oil.*

CHAPTER 21

1 *France Observateur*, Paris (10 August 1961).
2 See *Financial Times* (29 October 1962).
3 *Middle East Economic Survey*, Beirut (2 November 1962).
4 Article by Enrico Mattei in *Politique Étrangère*, Paris, 2 (1957).

CHAPTER 22

1 *Al Ahram*, Cairo (19 May 1961).
2 Rihani, *Ibn Saoud of Arabia.*

CHAPTER 23

1 In a conversation with the author.
2 In a conversation with the author.
3 Stocking, *Middle East Oil.*

4 *Al Ahram*, Cairo (2 December 1962).
5 *Al Hayah*, Beirut (23 December 1965).
6 *Middle East Economic Survey* (31 December 1965).
7 *Wall Street Journal* (8 February 1972).
8 *Wall Street Journal* (8 February 1972). Also quoted by *International Herald Tribune*, Paris (9 February 1972).
9 Pre-trial testimony of Guido Arata, quoted in the *Wall Street Journal* (8 February 1972).
10 *Wall Street Journal* (8 February 1972).
11 *Wall Street Journal* (8 February 1972).

CHAPTER 24

1 Information from a Saudi source who does not wish to be identified.
2 Statement issued by the Saudi government (13 June 1967).
3 Fuad Itayim, quoted in *Middle East Economic Survey supplement* (May 1970).
4 Quoted in *Middle East Economic Survey* (21 July 1967).
5 The account of Gaddafi's reactions is told by Sheikh Ahmad al Abah of Oman.
6 Statements in the *Sunday Times*, London (2 July 1967).
7 Statement at a press conference in London (9 August, 1967).
8 Report in *Al Siyasha*, Kuwait (15 April 1970).

CHAPTER 25

1 Quoted from the author's biography, *Curzon: The End of an Epoch* (London 1960).
2 Morris, James, *Sultan in Oman* (London 1957).
3 The various elements mixed up in these events, British and Omani alike, have been extremely discreet about them. This account comes from various sources, including the sultan himself as recounted to a number of his friends, including Dr Wendell Phillips.
4 The quotations come from *Le Monde*, Paris (27 May 1971).
5 Part of an announcement by BP Exploration Company (Libya) Limited in, among other newspapers, the *International Herald Tribune*, Paris (23 December 1971).

CHAPTER 26

1 In a conversation with the author.
2 Dana Adams Schmidt in the *New York Times* (28 August 1971).
3 A KOC official, in conversation with the author.
4 In a speech at the American University, Beirut (1969).
5 In a conversation with the author.

CHAPTER 27

1 In a conversation with the author.
2 In a conversation with the author.
3 Quoted from a paper presented at the third seminar on the Economics of the Petroleum Industry, held at the American University of Beirut (spring 1969).
4 Quoted from a paper presented at the third seminar on the Economics of the Petroleum Industry.
5 Quoted from a paper presented at the third seminar on the Economics of the Petroleum Industry.
6 Extract from President Bakr's speech on Baghdad Radio (1 June 1972), translated by Middle East Economic Survey, Beirut (2 June 1972).
7 In a conversation with the author.
8 In a conversation with the author.
9 In a conversation with the author.
10 Yamani, Ahmad, Economics of the Petroleum Industry.
11 Yamani, Ahmad, Economics of the Petroleum Industry.
12 In a conversation with the author.
13 In a conversation with the author.
14 In a conversation with the author.
15 In a conversation with the author.
16 In a conversation with the author.
17 In a conversation with the author.
18 An official of Aramco talking to the author.
19 Yamani, Ahmad, Economics of the Petroleum Industry.
20 Quoted from the Middle East Economic Survey, Beirut (10 March 1972.
21 Quoted from the Middle East Economic Survey, Beirut (10 March 1972.
22 Quoted by the official Saudi News Agency (SNA) from Jiddah (18 February 1972).
23 Quoted from the Middle East Economic Survey, Beirut (10 March 1972).
24 Quoted from the Middle East Economic Survey, Beirut (10 March 1972).
25 Quoted from the Middle East Economic Survey, Beirut (10 March 1972).

CHAPTER 28

1 Quoted in Time magazine (12 June 1972).
2 Bulletin Mensuel d' Information, Paris (25 October 1972).
3 Bulletin Mensuel d' Information, Paris (25 October 1972).
4 Time magazine (12 June 1972).
5 Faisal al-Mazadi in an article in the Kuwait Daily, al-Siyasah (15 November 1972).
6 Statement made at a press conference given in Paris (15 June

1972) following a meeting between King Hussein and President Pompidou.

7 Quoted from the *Bulletin de l'Industrie Pétrolière, Paris* (20 March 1972).

8 Quoted from the *Bulletin de l'Industrie Pétrolière, Paris* (20 March 1972).

9 Quoted from a lecture delivered by Dr Nadim Pachachi on 'The Role of OPEC in the Emergence of New Patterns in Government-Company Relationships' at the Royal Institute of International Affairs, Chatham House, London (19 May 1972).

10 Quoted in *Middle East Economic Survey* (16 June 1972).

11 Quoted in *Middle East Economic Survey* (16 June 1972).

Index

INDEX

Persia: owns Baku (until 1813), 2; oil in, 7, 14, 15; Russia and Britain in, 9, 13; *see also* Iran

Petrol rationing in Britain, end of, 131–2

Petroleum Development (Oman) Ltd, 289–90, 395

Petroleum Reserve Corporation, USA, 117

Philby, Harry St John, adviser to Ibn Saud, 38, 41, 43, 47; and Anglo-Persian, 49–50; and Socal, 50–3; on Saudi Arabian justice, 82; returns to Arabia (1946), 126–7; and Ibn Saud, 151, 152, 157

Phillips, Wendell, American archaeologist: in Dhofar, 198–200, 291; given oil concession; drills for oil unsuccessfully, 254–7

Phillips Oil, in Aminoil sundicate, 148

Philpryor Corporation, 199

Pipelines: in Baku oilfield, 3; from Mosul to Mediterranean, 25; from Persian Gulf to Mediterranean, *see* Tapline; from Dhahran to Bahrain undersea, 122, 127; in Libya, to Sirte, 279; from Iski to Gulf of Oman, 289–90; from South Israel to Mediterranean, 339n

Pirate Coast, Arabia, later Trucial States, 195, 295–6

Pompidou, President Georges, of France, 338

Popular Front for Liberation of Oman and the Arabian Gulf (PFLOAG), Communist organization, 290–1

"Posted price" for oil, 231–2; cut by Standard of New Jersey (1960), 232–3, 244, 261n; Libya and, 261, 278; cut by Aramco (1967), 325; would end with nationalization, 316–17, 343–4

Potter, Charley, drilling superintendent with Aramco, 105, 106

Power supplies from oil in Kuwait, 208, 311–12

Pretyman, E. G., M.P., and source of oil for British Navy, 12

Pritt, D. N., lawyer, 130

Pryor, Sam, vice-president of Pan American, 199

Pure Oil Company of Texas, 257

Qaboos, heir to throne of Oman, 291; succeeds to throne, 294

Qatar, Persian Gulf, 30, 61–2, 138; oil production in, stepped up to replace Iranian, 170, 179; drops out of proposed federation, 295

Radcliffe, Sir Cecil (later Lord), Gulbenkians' lawyer, 134, 135

Ras al Khaimah, Persian Gulf: drops out of proposed federation, 295; search for oil in, 296–8; Iran and, 298–9; joins federation, 299

Ras Tanura, oil port in Saudi Arabia, 84, 88, 326; lapses into disuse during war, 108–9; refinery at, 138

Rashid Ali al Gailani, leader of Golden Square rebellion, Iraq, 94, 97, 99; welcomed in Iran, 101

Razmara, General Ali, prime minister of Iran, 163

Red Line Agreement, 30–1, 46, 55; Gulbenkian's income from, 128, 129; breaking of, 130–8

Reynolds, G. B., field manager for D'Arcy, 9–10, 13, 14–15

Reza Khan, Shah-in-Shah of Persia, 92, 100; deposed and exiled by British and Russians, 103; body returned to Teheran, 159

Reza Pahlevi, Shah-in-Shah of Iran, 162–3, 175, 180–1, 212; and Tumb Island, 296, and OPEC, 273, 330, 333, gives interview on oil plans, 339–40

Rhoades, Ralph, American geologist, 46

Richfield Oil Corporation, 189n, 255

Rieber, Torkild, chairman of Texas Oil, 84

Rihani, Ameen, in Saudi Arabia, 32, 33, 34, 36, 37–8; on Arabian way of life, 78, 314

Riyadh, Saudi Arabian capital, 32, 33, 151, 152, 313–14; railway between Damman and, 153

Robertson, A. W. M., Anglo-Iranian manager at Kermanshah, 102–3

Rockefeller, John D., 3, 143, 243

Rodgers, W. S. S., of Texas Oil, 115–16, 118–19, 120

Roosevelt, President F. D., 111, 115, 116, 117; meets Ibn Saud (1944), 122–5

Rosenplaenter, Hiram, American oil-driller, 9, 10, 11

Ross, Cal, Aramco driller at Dhahran, 107n

Rothschilds, D'Arcy and, 71

Rouhani, Fouad, secretary and chairman of OPEC, 240

Rountree, William, under-secretary of US State Department, 224

"Rovin, General de" (F. F. L. Pegulu), and Libyan oil, 266, 269, 270

371